PYTHON DATA

STRUCTURES

Pocket Primer

PYTHON DATA STRUCTURES

Pocket Primer

Oswald Campesato

MERCURY LEARNING AND INFORMATION
Dulles, Virginia
Boston, Massachusetts
New Delhi

Publisher: David Pallai

MERCURY LEARNING AND INFORMATION
22841 Quicksilver Drive
Dulles, VA 20166
info@merclearning.com
www.merclearning.com
800-232-0223

O. Campesato. *Python Data Structures Pocket Primer.*
ISBN: 978-1-68392-757-0

The publisher recognizes and respects all marks used by companies, manufacturers, and developers as a means to distinguish their products. All brand names and product names mentioned in this book are trademarks or service marks of their respective companies. Any omission or misuse (of any kind) of service marks or trademarks, etc. is not an attempt to infringe on the property of others.

Library of Congress Control Number: 2022947023
222324321 This book is printed on acid-free paper in the United States of America.

Our titles are available for adoption, license, or bulk purchase by institutions, corporations, etc. For additional information, please contact the Customer Service Dept. at 800-232-0223(toll free).

All of our titles are available in digital format at academiccourseware.com and other digital vendors. Companion files (figures and code listings) for this title are available by contacting *info@merclearning.com*. The sole obligation of MERCURY LEARNING AND INFORMATION to the purchaser is to replace the disc, based on defective materials or faulty workmanship, but not based on the operation or functionality of the product.

I'd like to dedicate this book to my parents — may this bring joy and happiness into their lives.

CONTENTS

PREFACE

WHAT IS THE PRIMARY VALUE PROPOSITION FOR THIS BOOK?

This book contains a fast-paced introduction to as much relevant information about data structures that within reason can possibly be included in a book of this size. In addition, this book has a task-oriented approach, so you will see code samples that use data structures to solve various tasks.

Chapter 1 starts with an introduction to `Python` for beginners, recursion is discussed in Chapter 2, strings and arrays are covered in Chapter 3, search and sort algorithms are discussed in Chapter 4, various types of linked lists and explained in Chapter 5 and Chapter 6, and then queues and stacks are covered in Chapter 7.

Please keep in mind that a full treatment of all the topics in this book could easily triple the length of this book (and besides, such books are already available).

THE TARGET AUDIENCE

This book is intended primarily for people who have a limited background in data structures. This book is also intended to reach an international audience of readers with highly diverse backgrounds. While many readers know how to read English, their native spoken language is not English (which could be their second, third, or even fourth language). Consequently, this book uses standard English rather than colloquial expressions in order to maximize clarity.

WHAT WILL I LEARN FROM THIS BOOK?

The introductory section of the preface contains a brief outline of the topics in each of the chapters of this book. As the title suggests, you will acquire a basic level of knowledge about a variety of data structures.

Incidentally, you will discover that many code samples contain "commented out" code snippets, which are usually `Python print()` statements. Feel free to "uncomment" those code snippets, which will enable you to see the various execution paths in the code. In essence, you will see the debugging process that was used during the development of the code samples.

WHY ARE THE CODE SAMPLES PRIMARILY IN PYTHON?

Most of the code samples are short (usually less than one page and sometimes less than half a page), and if necessary, you can easily and quickly copy/paste the code into a new Jupyter notebook. Moreover, the code samples execute quickly, so you won't need to avail yourself of the free GPU that is provided in Google Colaboratory.

DO I NEED TO LEARN THE THEORY PORTIONS OF THIS BOOK?

Alas, an understanding of the theoretical underpinnings of data structures does not translate into the ability to solve tasks involving data structures: it's necessary knowledge (but not necessarily sufficient). Strive for an understanding of concepts and minimize the amount of memorization of code samples. For example, you can determine whether or not a positive integer n is a power of 2 with a single line of code:

```
(n > 0) and (n & (n-1)) == 0
```

Although the preceding code snippet might seem nonintuitive, you can convince yourself that this is true by setting n=8 and then observe the following:

```
n:   1000
n-1: 0111
```

The key point is this: the binary representation of a power of 2 has a single 1 in the left-most position, and zeroes to the right of the digit 1 (for n>=2), whereas the number n-1 contains a 0 in the left-most position, and all 1s to the right of the digit 0. Therefore, the logical and of n and (n-1) is clearly 0.

Now that you understand the key idea, there is no need to memorize anything, and you can write the solution in any programming language for which you have a very modest level of experience.

The theoretical aspects will help you improve your conceptual understanding of the differences and similarities (if any) among various types of data structures. However, you will gain confidence and also a better understanding of data structures by writing code because knowledge is often gained through repetition of tasks that provide reinforcement of concepts.

GETTING THE MOST FROM THIS BOOK

Some programmers learn well from prose, others learn well from sample code (and lots of it), which means that there's no single style that can be used for everyone.

Moreover, some programmers want to run the code first, see what it does, and then return to the code to delve into the details (and others use the opposite approach).

Consequently, there are various types of code samples in this book: some are short, some are long, and other code samples "build" from earlier code samples.

WHAT DO I NEED TO KNOW FOR THIS BOOK?

Current knowledge of Python 3.x is useful because all the code samples are in `Python`. Knowledge of data structures will enable you to progress through the related chapters more quickly. The less technical knowledge you have, the more diligence will be required in order to understand the various topics that are covered.

If you want to be sure that you can grasp the material in this book, glance through some of the code samples to get an idea of how much is familiar to you and how much is new for you.

DOES THIS BOOK CONTAIN PRODUCTION-LEVEL CODE SAMPLES?

The primary purpose of the code samples in this book is to show you solutions to tasks that involve data structures. Therefore, clarity has higher priority than writing more compact or highly optimized code, For example, inspect the loops in the Python code sample to see if they can be made more efficient. Suggestion: treat such code samples as opportunities for you to optimize the code samples in this book.

If you decide to use any of the code in this book in a production website, you ought to subject that code to the same rigorous analysis as the other parts of your code base.

WHAT ARE THE NONTECHNICAL PREREQUISITES FOR THIS BOOK?

Although the answer to this question is difficult to quantify, it's especially important to have a strong desire to learn about data analytics, along with the motivation and discipline to read and understand the code samples.

HOW DO I SET UP A COMMAND SHELL?

If you are a Mac user, there are three ways to set up a command shell. The first method is to use Finder to navigate to Applications > Utilities and then double click on the Utilities application. Next, if you already have a command shell available, you can launch a new command shell by typing the following command:

```
open /Applications/Utilities/Terminal.app
```

A second method for Mac users is to open a new command shell on a MacBook from a command shell that is already visible simply by clicking command+n in that command shell, and your Mac will launch another command shell.

If you are a PC user, you can install Cygwin (open source *https://cygwin.com/*) that simulates bash commands, or use another toolkit such as MKS (a commercial product). Please read the online documentation that describes the download and installation process. Note that custom aliases are not automatically set if they are defined in a file other than the main start-up file (such as .bash_login).

COMPANION FILES

All the code samples in this book may be obtained by writing to the publisher at *info@merclearning.com*.

WHAT ARE THE "NEXT STEPS" AFTER FINISHING THIS BOOK?

The answer to this question varies widely, mainly because the answer depends heavily on your objectives. If you are interested primarily in learning more about data structures, then this book is a "stepping stone" to other books that contain more complex data structures as well as code samples for the tasks that are not covered in this book (such as deleting a node from a tree or a doubly linked list).

If you are primarily interested in machine learning, there are some subfields of machine learning, such as deep learning and reinforcement learning (and deep reinforcement learning) that might appeal to you. Fortunately, there are many resources available, and you can perform an Internet search for those resources. One other point: the aspects of machine learning for you to learn depend on who you are—the needs of a machine learning engineer, data scientist, manager, student, or software developer all differ from one another.

INTRODUCTION TO PYTHON

This chapter provides an introduction to basic features of Python, including examples of working with Python strings, arrays, and dictionaries. Please keep in mind that this chapter does not contain details about the Python interpreter: you can find that information online in various tutorials.

You will also learn about useful tools for installing Python modules, basic Python constructs, and how to work with some data types in Python.

The first part of this chapter shows you how to work with simple data types, such as numbers, fractions, and strings. The third part of this chapter discusses exceptions and how to use them in Python scripts.

The second part of this chapter introduces you to various ways to perform conditional logic in Python, as well as control structures and user-defined functions in Python. Virtually every Python program that performs useful calculations requires some type of conditional logic or control structure (or both). Although the syntax for these Python features is slightly different from other languages, the functionality will be familiar to you.

The third part of this chapter contains examples that involve nested loops and user-defined Python functions. The remaining portion of the chapter discusses tuples, sets, and dictionaries.

NOTE *The* Python *scripts in this book are for* Python *3.x.*

SOME STANDARD MODULES IN PYTHON

The Python Standard Library provides many modules that can simplify your own Python scripts. A list of the Standard Library modules is here:

http://www.python.org/doc/

Some of the most important `Python` modules include `cgi`, `math`, `os`, `pickle`, `random`, `re`, `socket`, `sys`, `time`, and `urllib`.

The code samples in this book use the modules `math`, `os`, `random`, and `re`. You need to import these modules in order to use them in your code. For example, the following code block shows you how to import standard `Python` modules:

```
import re
import sys
import time
```

The code samples in this book import one or more of the preceding modules, as well as other `Python` modules. The next section discusses primitive data types in `Python`.

SIMPLE DATA TYPES IN PYTHON

`Python` supports primitive data types, such as numbers (integers, floating point numbers, and exponential numbers), strings, and dates. `Python` also supports more complex data types, such as lists (or arrays), tuples, and dictionaries, all of which are discussed later in this chapter. The next several sections discuss some of the `Python` primitive data types, along with code snippets that show you how to perform various operations on those data types.

WORKING WITH NUMBERS

`Python` provides arithmetic operations for manipulating numbers a straightforward manner that is similar to other programming languages. The following examples involve arithmetic operations on integers:

```
>>> 2+2
4
>>> 4/3
1
>>> 3*8
24
```

The following example assigns numbers to two variables and computes their product:

```
>>> x = 4
>>> y = 7
>>> x * y
28
```

The following examples demonstrate arithmetic operations involving integers:

```
>>> 2+2
4
```

```
>>> 4/3
1
>>> 3*8
24
```

Notice that division ("/") of two integers is actually truncation in which only the integer result is retained. The following example converts a floating point number into exponential form:

```
>>> fnum = 0.00012345689000007
>>> "%.14e"%fnum
'1.23456890000070e-04'
```

You can use the int() function and the float() function to convert strings to numbers:

```
word1 = "123"
word2 = "456.78"
var1 = int(word1)
var2 = float(word2)
print("var1: ",var1," var2: ",var2)
```

The output from the preceding code block is here:

```
var1:   123   var2:   456.78
```

Alternatively, you can use the eval() function:

```
word1 = "123"
word2 = "456.78"
var1 = eval(word1)
var2 = eval(word2)
print("var1: ",var1," var2: ",var2)
```

If you attempt to convert a string that is not a valid integer or a floating point number, Python raises an exception, so it's advisable to place your code in a try/except block (discussed later in this chapter).

Working With Other Bases

Numbers in Python are in base 10 (the default), but you can easily convert numbers to other bases. For example, the following code block initializes the variable x with the value 1234, and then displays that number in base 2, 8, and 16, respectively:

```
>>> x = 1234
>>> bin(x) '0b10011010010'
>>> oct(x) '0o2322'
>>> hex(x) '0x4d2'
```

Use the `format()` function if you want to suppress the 0b, 0o, or 0x prefixes, as shown here:

```
>>> format(x, 'b')  '10011010010'
>>> format(x, 'o')  '2322'
>>> format(x, 'x')  '4d2'
```

Negative integers are displayed with a negative sign:

```
>>> x = -1234
>>> format(x, 'b')  '-10011010010'
>>> format(x, 'x')  '-4d2'
```

The chr() **Function**

The Python `chr()` function takes a positive integer as a parameter and converts it to its corresponding alphabetic value (if one exists). The letters A through Z have decimal representation of 65 through 91 (which corresponds to hexadecimal 41 through 5b), and the lowercase letters a through z have decimal representation 97 through 122 (hexadecimal 61 through 7b).

Here is an example of using the `chr()` function to print uppercase A:

```
>>> x=chr(65)
>>> x
'A'
```

The following code block prints the ASCII values for a range of integers:

```
result = ""
for x in range(65,90):
  print(x, chr(x))
  result = result+chr(x)+' '
print("result: ",result)
```

NOTE Python 2 *uses* ASCII *strings whereas* Python 3 *uses* UTF-8.

You can represent a range of characters with the following line:

```
for x in range(65,90):
```

However, the following equivalent code snippet is more intuitive:

```
for x in range(ord('A'), ord('Z')):
```

If you want to display the result for lowercase letters, change the preceding range from (65,91) to either of the following statements:

```
for x in range(65,90):
for x in range(ord('a'), ord('z')):
```

The `round()` **Function in Python**

The `Python round()` function enables you to round decimal values to the nearest precision:

```
>>> round(1.23, 1)
1.2
>>> round(-3.42,1)
-3.4
```

Before delving into `Python` code samples that work with strings, the next section briefly discusses `Unicode` and `UTF-8`, both of which are character encodings.

UNICODE AND UTF-8

A `Unicode` string consists of a sequence of numbers that are between `0` and `0x10ffff`, where each number represents a group of bytes. An encoding is the manner in which a `Unicode` string is translated into a sequence of bytes. Among the various encodings, `UTF-8` (Unicode Transformation Format) is perhaps the most common, and it's also the default encoding for many systems. The digit 8 in `UTF-8` indicates that the encoding uses 8-bit numbers, whereas `UTF-16` uses 16-bit numbers (but this encoding is less common).

The `ASCII` character set is a subset of `UTF-8`, so a valid `ASCII` string can be read as a `UTF-8` string without any re-encoding required. In addition, a `Unicode` string can be converted into a `UTF-8` string.

WORKING WITH UNICODE

`Python` supports `Unicode`, which means that you can render characters in different languages. `Unicode` data can be stored and manipulated in the same way as strings. Create a `Unicode` string by prepending the letter "u," as shown here:

```
>>> u'Hello from Python!'
u'Hello from Python!'
```

Special characters can be included in a string by specifying their `Unicode` value. For example, the following `Unicode` string embeds a space (which has the `Unicode` value 0x0020) in a string:

```
>>> u'Hello\u0020from Python!'
u'Hello from Python!'
```

Listing 1.1 displays the contents of `Unicode1.py` that illustrates how to display a string of characters in Japanese and another string of characters in Chinese (Mandarin).

LISTING 1.1: Unicode1.py

```
chinese1 = u'\u5c07\u63a2\u8a0e HTML5 \u53ca\u5176\u4ed6'
hiragana = u'D3 \u306F \u304B\u3063\u3053\u3043\u3043 \u3067\u3059!'

print('Chinese:',chinese1)
print('Hiragana:',hiragana)
```

The output of Listing 1.1 is here:

```
Chinese: 將探討 HTML5 及其他
Hiragana: D3 は かっこいい です!
```

The next portion of this chapter shows you how to "slice and dice" text strings with built-in Python functions.

WORKING WITH STRINGS

Literal strings in Python 3 are Unicode by default. You can concatenate two strings using the '+' operator. The following example prints a string and then concatenates two single-letter strings:

```
>>> 'abc'
'abc'
>>> 'a' + 'b'
'ab'
```

You can use '+' or '*' to concatenate identical strings, as shown here:

```
>>> 'a' + 'a' + 'a'
'aaa'
>>> 'a' * 3
'aaa'
```

You can assign strings to variables and print them using the print() statement:

```
>>> print('abc')
abc
>>> x = 'abc'
>>> print(x)
abc
>>> y = 'def'
>>> print(x + y)
abcdef
```

You can "unpack" the letters of a string and assign them to variables, as shown here:

```
>>> str = "World"
>>> x1,x2,x3,x4,x5 = str
>>> x1
'W'
```

```
>>> x2
'o'
>>> x3
'r'
>>> x4
'l'
>>> x5
'd'
```

The preceding code snippets shows you how easy it is to extract the letters in a text string. You can also extract substrings of a string as shown in the following examples:

```
>>> x = "abcdef"
>>> x[0]
'a'
>>> x[-1]
'f'
>>> x[1:3]
'bc'
>>> x[0:2] + x[5:]
'abf'
```

However, you will cause an error if you attempt to subtract two strings, as you probably expect:

```
>>> 'a' - 'b'
Traceback (most recent call last):
  File "<stdin>", line 1, in <module>
TypeError: unsupported operand type(s) for -: 'str' and 'str'
```

The try/except construct in Python (discussed later in this chapter) enables you to handle the preceding type of exception more gracefully.

Comparing Strings

You can use the methods lower() and upper() to convert a string to lowercase and uppercase, respectively, as shown here:

```
>>> 'Python'.lower()
'python'
>>> 'Python'.upper()
'PYTHON'
>>>
```

The methods lower() and upper() are useful for performing a case insensitive comparison of two ASCII strings. Listing 1.2 displays the contents of Compare. py that uses the lower() function in order to compare two ASCII strings.

LISTING 1.2: Compare.py

```
x = 'Abc'
y = 'abc'
```

```
if(x == y):
  print('x and y: identical')
elif (x.lower() == y.lower()):
  print('x and y: case insensitive match')
else:
  print('x and y: different')
```

Since x contains mixed case letters and y contains lowercase letters, Listing 1.2 displays the following output:

```
x and y: different
```

Uninitialized Variables and the Value None in Python

Python distinguishes between an uninitialized variable and the value None. The former is a variable that has not been assigned a value, whereas the value None is a value that indicates "no value." Collections and methods often return the value None, and you can test for the value None in conditional logic (shown later in this chapter).

The next portion of this chapter shows you how to "slice and dice" text strings with built-in Python functions.

SLICING AND SPLICING STRINGS

Python enables you to extract substrings of a string (called "slicing") using array notation. Slice notation is start:stop:step, where the start, stop, and step values are integers that specify the start value, end value, and the increment value. The interesting part about slicing in Python is that you can use the value -1, which operates from the right side instead of the left side of a string. Some examples of slicing a string are here:

```
text1 = "this is a string"
print('First 7 characters:',text1[0:7])
print('Characters 2-4:',text1[2:4])
print('Right-most character:',text1[-1])
print('Right-most 2 characters:',text1[-3:-1])
```

The output from the preceding code block is here:

```
First 7 characters: this is
Characters 2-4: is
Right-most character: g
Right-most 2 characters: in
```

Later in this chapter you will see how to insert a string in the middle of another string.

Testing for Digits and Alphabetic Characters

Python enables you to examine each character in a string and then test whether that character is a bona fide digit or an alphabetic character.

This section provides a precursor to regular expressions that are discussed in Chapter 8.

Listing 1.3 displays the contents of CharTypes.py that illustrates how to determine if a string contains digits or characters. In case you are unfamiliar with the conditional "if" statement in Listing 1.3, more detailed information is available later in this chapter.

LISTING 1.3: CharTypes.py

```
str1 = "4"
str2 = "4234"
str3 = "b"
str4 = "abc"
str5 = "a1b2c3"

if(str1.isdigit()):
  print("this is a digit:",str1)

if(str2.isdigit()):
  print("this is a digit:",str2)

if(str3.isalpha()):
  print("this is alphabetic:",str3)

if(str4.isalpha()):
  print("this is alphabetic:",str4)

if(not str5.isalpha()):
  print("this is not pure alphabetic:",str5)

print("capitalized first letter:",str5.title())
```

Listing 1.3 initializes some variables, followed by two conditional tests that check whether or not str1 and str2 are digits using the isdigit() function. The next portion of Listing 1.3 checks if str3, str4, and str5 are alphabetic strings using the isalpha() function. The output of Listing 1.3 is here:

```
this is a digit: 4
this is a digit: 4234
this is alphabetic: b
this is alphabetic: abc
this is not pure alphabetic: a1b2c3
capitalized first letter: A1B2C3
```

SEARCH AND REPLACE A STRING IN OTHER STRINGS

Python provides methods for searching and also for replacing a string in a second text string. Listing 1.4 displays the contents of FindPos1.py that shows you how to use the find() function to search for the occurrence of one string in another string.

LISTING 1.4 FindPos1.py

```
item1 = 'abc'
item2 = 'Abc'
text = 'This is a text string with abc'

pos1 = text.find(item1)
pos2 = text.find(item2)

print('pos1=',pos1)
print('pos2=',pos2)
```

Listing 1.4 initializes the variables item1, item2, and text, and then searches for the index of the contents of item1 and item2 in the string text. The Python find() function returns the column number where the first successful match occurs; otherwise, the find() function returns a −1 if a match is unsuccessful. The output from launching Listing 1.4 is here:

```
pos1= 27
pos2= -1
```

In addition to the find() method, you can use the in operator when you want to test for the presence of an element, as shown here:

```
>>> lst = [1,2,3]
>>> 1 in lst
True
```

Listing 1.5 displays the contents of Replace1.py that shows you how to replace one string with another string.

LISTING 1.5: Replace1.py

```
text = 'This is a text string with abc'
print('text:',text)
text = text.replace('is a', 'was a')
print('text:',text)
```

Listing 1.5 starts by initializing the variable text and then printing its contents. The next portion of Listing 1.5 replaces the occurrence of "is a" with "was a" in the string text, and then prints the modified string. The output from launching Listing 1.5 is here:

```
text: This is a text string with abc
text: This was a text string with abc
```

PRECEDENCE OF OPERATORS IN PYTHON

When you have an expression involving numbers, you might remember that multiplication ("*") and division ("/") have higher precedence than addition ("+") or subtraction ("−"). Exponentiation has even higher precedence than these four arithmetic operators.

However, instead of relying on precedence rules, it's simpler (as well as safer) to use parentheses. For example, `(x/y)+10` is clearer than `x/y+10`, even though they are equivalent expressions.

As another example, the following two arithmetic expressions are the equivalent, but the second is less error prone than the first:

```
x/y+3*z/8+x*y/z-3*x
(x/y)+(3*z)/8+(x*y)/z-(3*x)
```

In any case, the following website contains precedence rules for operators in `Python`:

http://www.mathcs.emory.edu/~valerie/courses/fall10/155/resources/ op_precedence.html

PYTHON RESERVED WORDS

Every programming language has a set of reserved words, which is a set of words that cannot be used as identifiers, and `Python` is no exception. The `Python` reserved words are: and, exec, not, assert, finally, or, break, for, pass, class, from, print, continue, global, raise, def, if, return, del, import, try, elif, in, while, else, is, with, except, lambda, and yield.

If you inadvertently use a reserved word as a variable, you will see an "invalid syntax" error message instead of a "reserved word" error message. For example, suppose you create a `Python` script `test1.py` with the following code:

```
break = 2
print('break =', break)
```

If you run the preceding Python code you will see the following output:

```
File "test1.py", line 2
    break = 2
          ^
SyntaxError: invalid syntax
```

However, a quick inspection of the `Python` code reveals the fact that you are attempting to use the reserved word `break` as a variable.

WORKING WITH LOOPS IN PYTHON

Python supports for loops, while loops, and range() statements. The following subsections illustrate how you can use each of these constructs.

Python for Loops

Python supports the for loop whose syntax is slightly different from other languages (such as JavaScript and Java). The following code block shows you how to use a for loop in Python in order to iterate through the elements in a list:

```
>>> x = ['a', 'b', 'c']
>>> for w in x:
...     print(w)
...
a
b
c
```

The preceding code snippet prints three letters on three separate lines. You can force the output to be displayed on the same line (which will "wrap" if you specify a large enough number of characters) by appending a comma "," in the print() statement, as shown here:

```
>>> x = ['a', 'b', 'c']
>>> for w in x:
...     print(w, end=' ')
...
a b c
```

You can use this type of code when you want to display the contents of a text file in a single line instead of multiple lines.

Python also provides the built-in reversed() function that reverses the direction of the loop, as shown here:

```
>>> a = [1, 2, 3, 4, 5]
>>> for x in reversed(a):
... print(x)
5
4
3
2
1
```

Note that reversed iteration only works if the size of the current object can be determined or if the object implements a __reversed__() special method.

Numeric Exponents in Python

Listing 1.6 displays the contents of Nth_exponent.py that illustrates how to calculate intermediate powers of a set of integers.

LISTING 1.6: Nth_exponent.py

```
maxPower = 4
maxCount = 4

def pwr(num):
  prod = 1
  for n in range(1,maxPower+1):
    prod = prod*num
    print(num,'to the power',n, 'equals',prod)
  print('-----------')

for num in range(1,maxCount+1):
    pwr(num)
```

Listing 1.6 contains a function called `pwr()` that accepts a numeric value. This function contains a loop that prints the value of that number raised to the power n, where n ranges between `1` and `maxPower+1`.

The second part of Listing 1.6 contains a `for` loop that invokes the function `pwr()` with the numbers between `1` and `maxPower+1`. The output from Listing 1.16 is here:

```
1 to the power 1 equals 1
1 to the power 2 equals 1
1 to the power 3 equals 1
1 to the power 4 equals 1
-----------
2 to the power 1 equals 2
2 to the power 2 equals 4
2 to the power 3 equals 8
2 to the power 4 equals 16
-----------
3 to the power 1 equals 3
3 to the power 2 equals 9
3 to the power 3 equals 27
3 to the power 4 equals 81
-----------
4 to the power 1 equals 4
4 to the power 2 equals 16
4 to the power 3 equals 64
4 to the power 4 equals 256
-----------
```

NESTED LOOPS

Listing 1.7 displays the contents of `Triangular1.py` that illustrates how to print a row of consecutive integers (starting from `1`), where the length of each row is one greater than the previous row.

LISTING 1.7: Triangular1.py

```
max = 8
for x in range(1,max+1):
```

```
    for y in range(1,x+1):
      print(y, '', end='')
    print()
```

Listing 1.7 initializes the variable max with the value 8, followed by an outer for loop whose loop variable x ranges from 1 to max+1. The inner loop has a loop variable y that ranges from 1 to x+1, and the inner loop prints the value of y. The output of Listing 1.7 is here:

```
1
1 2
1 2 3
1 2 3 4
1 2 3 4 5
1 2 3 4 5 6
1 2 3 4 5 6 7
1 2 3 4 5 6 7 8
```

THE SPLIT() FUNCTION WITH FOR LOOPS

Python supports various useful string-related functions, including the split() function and the join() function. The split() function is useful when you want to tokenize ("split") a line of text into words and then use a for loop to iterate through those words and process them accordingly.

The join() function does the opposite of split(): it "joins" two or more words into a single line. You can easily remove extra spaces in a sentence by using the split() function and then invoking the join() function, thereby creating a line of text with one white space between any two words.

USING THE SPLIT() FUNCTION TO COMPARE WORDS

Listing 1.8 displays the contents of Compare2.py that illustrates how to use the split function to compare each word in a text string with another word.

LISTING 1.8: Compare2.py

```
x = 'This is a string that contains abc and Abc'
y = 'abc'
identical = 0
casematch = 0

for w in x.split():
  if(w == y):
    identical = identical + 1
  elif (w.lower() == y.lower()):
    casematch = casematch + 1

if(identical > 0):
 print('found identical matches:', identical)
```

```
if(casematch > 0):
 print('found case matches:', casematch)

if(casematch == 0 and identical == 0):
 print('no matches found')
```

Listing 1.8 uses the split() function in order to compare each word in the string x with the word abc. If there is an exact match, the variable identical is incremented. If a match does not occur, a case-insensitive match of the current word is performed with the string abc, and the variable casematch is incremented if the match is successful. The output from Listing 1.8 is here:

```
found identical matches: 1
found case matches: 1
```

PYTHON WHILE LOOPS

You can define a while loop to iterate through a set of numbers, as shown in the following examples:

```
>>> x = 0
>>> while x < 5:
...     print(x)
...     x = x + 1
...
0
1
2
3
4
5
```

Python uses indentation instead of curly braces that are used in other languages such as JavaScript and Java. Although the Python list data structure is not discussed until later in this chapter, you can probably understand the following simple code block that contains a variant of the preceding while loop that you can use when working with lists:

```
lst  = [1,2,3,4]

while lst:
  print('list:',lst)
  print('item:',lst.pop())
```

The preceding while loop terminates when the lst variable is empty, and there is no need to explicitly test for an empty list. The output from the preceding code is here:

```
list: [1, 2, 3, 4]
item: 4
```

```
list: [1, 2, 3]
item: 3
list: [1, 2]
item: 2
list: [1]
item: 1
```

This concludes the examples that use the `split()` function in order to process words and characters in a text string. The next part of this chapter shows you examples of using conditional logic in `Python` code.

CONDITIONAL LOGIC IN PYTHON

If you have written code in other programming languages, you have undoubtedly seen `if/then/else` (or `if-elseif-else`) conditional statements. Although the syntax varies between languages, the logic is essentially the same. The following example shows you how to use `if/elif` statements in `Python`:

```
>>> x = 25
>>> if x < 0:
...    print('negative')
... elif x < 25:
...    print('under 25')
... elif x == 25:
...    print('exactly 25')
... else:
...    print('over 25')
...
exactly 25
```

The preceding code block illustrates how to use multiple conditional statements, and the output is exactly what you expected.

THE BREAK/CONTINUE/PASS STATEMENTS

The `break` statement in `Python` enables you to perform an "early exit" from a loop, whereas the `continue` statement essentially returns to the top of the loop and continues with the next value of the loop variable. The `pass` statement is essentially a "do nothing" statement.

Listing 1.9 displays the contents of `BreakContinuePass.py` that illustrates the use of these three statements.

LISTING 1.9: BreakContinuePass.py

```
print('first loop')
for x in range(1,4):
  if(x == 2):
    break
  print(x)
```

```
print('second loop')
for x in range(1,4):
  if(x == 2):
    continue
  print(x)

print('third loop')
for x in range(1,4):
  if(x == 2):
    pass
  print(x)
```

The output of Listing 1.9 is here:

```
first loop
1
second loop
1
3
third loop
1
2
3
```

COMPARISON AND BOOLEAN OPERATORS

Python supports a variety of Boolean operators, such as in, not in, is, is not, and, or, and not. The next several sections discuss these operators and provide some examples of how to use them.

The in/not in/is/is not **Comparison Operators**

The in and not in operators are used with sequences to check whether a value occurs or does not occur in a sequence. The operators is and is not determine whether or not two objects are the same object, which is important only matters for mutable objects such as lists. All comparison operators have the same priority, which is lower than that of all numerical operators. Comparisons can also be chained. For example, a < b == c tests whether a is less than b and moreover b equals c.

The and, or, and not Boolean **Operators**

The Boolean operators and, or, and not have lower priority than comparison operators. The Boolean and and or are binary operators whereas the Boolean or operator is a unary operator. Here are some examples:

- A and B can only be true if both A and B are true
- A or B is true if either A or B is true
- not(A) is true if and only if A is false

You can also assign the result of a comparison or other `Boolean` expression to a variable, as shown here:

```
>>> string1, string2, string3 = '', 'b', 'cd'
>>> str4 = string1 or string2 or string3
>>> str4
'b'
```

The preceding code block initializes the variables `string1`, `string2`, and `string3`, where `string1` is an empty string. Next, `str4` is initialized via the `or` operator, and since the first nonnull value is `string2`, the value of `str4` is equal to `string2`.

LOCAL AND GLOBAL VARIABLES

`Python` variables can be local or global. A `Python` variable is local to a function if the following are true:

- a parameter of the function
- on the left side of a statement in the function
- bound to a control structure (such as for, with, and except)

A variable that is referenced in a function but is not local (according to the previous list) is a non-local variable. You can specify a variable as nonlocal with this snippet:

```
nonlocal z
```

A variable can be explicitly declared as global with this statement:

```
global z
```

The following code block illustrates the behavior of a global versus a local variable:

```
global z
z = 3

def changeVar(z):
  z = 4
  print('z in function:',z)

print('first global z:',z)

if __name__ == '__main__':
  changeVar(z)
  print('second global z:',z)
```

The output from the preceding code block is here:

```
first global z: 3
z in function: 4
second global z: 3
```

SCOPE OF VARIABLES

The accessibility or scope of a variable depends on where that variable has been defined. Python provides two scopes: global and local, with the added "twist" that global is actually module-level scope (i.e., the current file), and therefore you can have a variable with the same name in different files and they will be treated differently.

Local variables are straightforward: they are defined inside a function, and they can only be accessed inside the function where they are defined. Any variables that are not local variables have global scope, which means that those variables are "global" *only* with respect to the file where it has been defined, and they can be accessed anywhere in a file.

There are two scenarios to consider regarding variables. First, suppose two files (aka modules) file1.py and file2.py have a variable called x, and file1. py also imports file2.py. The question now is how to disambiguate between the x in the two different modules. As an example, suppose that file2.py contains the following two lines of code:

```
x = 3
print('unscoped x in file2:',x)
```

Suppose that file1.py contains the following code:

```
import file2 as file2

x = 5
print('unscoped x in file1:',x)
print('scoped x from file2:',file2.x)
```

Launch file1.py from the command line, and you will see the following output:

```
unscoped x in file2: 3
unscoped x in file1: 5
scoped x from file2: 3
```

The second scenario involves a program contains a local variable and a global variable with the same name. According to the earlier rule, the local variable is used in the function where it is defined, and the global variable is used outside of that function.

The following code block illustrates the use of a global and local variable with the same name:

```
#!/usr/bin/python
# a global variable:
total = 0;

def sum(x1, x2):
    # this total is local:
    total = x1+x2;
```

```
    print("Local total : ", total)
    return total

# invoke the sum function
sum(2,3);
print("Global total : ", total)
```

When the above code is executed, it produces following result:

```
Local total :    5
Global total :   0
```

What about unscoped variables, such as specifying the variable x without a module prefix? The answer consists of the following sequence of steps that Python will perform:

1. Check the local scope for the name.
2. Ascend the enclosing scopes and check for the name.
3. Perform step 2 until the global scope (i.e., module level).
4. If x still hasn't been found, Python checks __builtins__.

```
Python 3.9.1 (v3.9.1:1e5d33e9b9, Dec  7 2020, 12:44:01)
[Clang 12.0.0 (clang-1200.0.32.27)] on darwin
Type "help", "copyright", "credits" or "license" for more information.
>>> x = 1
>>> g = globals()
>>> g
{'g': {...}, '__builtins__': <module '__builtin__' (built-in)>, '__
package__': None, 'x': 1, '__name__': '__main__', '__doc__': None}
>>> g.pop('x')
1
>>> x
Traceback (most recent call last):
  File "<stdin>", line 1, in <module>
NameError: name 'x' is not defined
```

> **NOTE** *You can access the* dicts *that* Python *uses to track local and global scope by invoking* locals() *and* globals() *respectively.*

PASS BY REFERENCE VERSUS VALUE

All parameters (arguments) in the Python language are passed by reference. Thus, if you change what a parameter refers to within a function, the change is reflected in the calling function. For example:

```
def changeme(mylist):
    #This changes a passed list into this function
    mylist.append([1,2,3,4])
    print("Values inside the function: ", mylist)
    return
```

```
# Now you can call changeme function
mylist = [10,20,30]
changeme(mylist)
print("Values outside the function: ", mylist)
```

Here we are maintaining reference of the passed object and appending values in the same object, and the result is shown here:

```
Values inside the function:  [10, 20, 30, [1, 2, 3, 4]]
Values outside the function:  [10, 20, 30, [1, 2, 3, 4]]
```

The fact that values are passed by reference gives rise to the notion of mutability versus immutability that is discussed in Chapter 3.

ARGUMENTS AND PARAMETERS

Python differentiates between arguments to functions and parameter declarations in functions: a positional (mandatory) and keyword (optional/default value). This concept is important because Python has operators for packing and unpacking these kinds of arguments. Python unpacks positional arguments from an iterable, as shown here:

```
>>> def foo(x, y):
...    return x - y
...
>>> data = 4,5
>>> foo(data) # only passed one arg
Traceback (most recent call last):
  File "<stdin>", line 1, in <module>
TypeError: foo() takes exactly 2 arguments (1 given)
>>> foo(*data) # passed however many args are in tuple
-1
```

USER-DEFINED FUNCTIONS IN PYTHON

Python provides built-in functions and also enables you to define your own functions. You can define functions to provide the required functionality. Here are simple rules to define a function in Python:

- Function blocks begin with the keyword def followed by the function name and parentheses.
- Any input arguments should be placed within these parentheses.
- The first statement of a function can be an optional statement—the documentation string of the function or docstring.
- The code block within every function starts with a colon (:) and is indented.
- The statement return [expression] exits a function, optionally passing back an expression to the caller. A return statement with no arguments is the same as return "None."
- If a function does not specify return statement, the function automatically returns "None," which is a special type of value in Python.

A very simple custom `Python` function is here:

```
>>> def func():
...     print 3
...
>>> func()
3
```

The preceding function is trivial, but it does illustrate the syntax for defining custom functions in `Python`. The following example is slightly more useful:

```
>>> def func(x):
...     for i in range(0,x):
...         print(i)
...
>>> func(5)
0
1
2
3
4
```

SPECIFYING DEFAULT VALUES IN A FUNCTION

Listing 1.10 displays the contents of `DefaultValues.py` that illustrates how to specify default values in a function.

LISTING 1.10: DefaultValues.py

```
def numberFunc(a, b=10):
  print (a,b)

def stringFunc(a, b='xyz'):
  print (a,b)

def collectionFunc(a, b=None):
  if(b is None):
     print('No value assigned to b')

numberFunc(3)
stringFunc('one')
collectionFunc([1,2,3])
```

Listing 1.10 defines three functions, followed by an invocation of each of those functions. The functions `numberFunc()` and `stringFunc()` print a list contain the values of their two parameters, and `collection-Func()` displays a message if the second parameter is `None`. The output from Listing 1.10 is here:

```
(3, 10)
('one', 'xyz')
No value assigned to b
```

Returning Multiple Values From a Function

This task is accomplished by the code in Listing 1.11, which displays the contents of `MultipleValues.py`.

LISTING 1.11: MultipleValues.py

```
def MultipleValues():
    return 'a', 'b', 'c'

x, y, z = MultipleValues()

print('x:',x)
print('y:',y)
print('z:',z)
```

The output from Listing 1.11 is here:

```
x: a
y: b
z: c
```

LAMBDA EXPRESSIONS

Listing 1.12 displays the contents of `Lambda1.py` that illustrates how to create a simple lambda function in `Python`.

LISTING 1.12: Lambda1.py

```
add = lambda x, y: x + y

x1 = add(5,7)
x2 = add('Hello', 'Python')

print(x1)
print(x2)
```

Listing 1.12 defines the lambda expression add that accepts two input parameters and then returns their sum (for numbers) or their concatenation (for strings).

The output from Listing 1.12 is here:

```
12
HelloPython
```

The next portion of this chapter discusses `Python` collections, such as lists (or arrays), sets, tuples, and dictionaries. You will see many short code blocks that will help you rapidly learn how to work with these data structures in `Python`. After you have finished reading this chapter, you will be in a better position to create more complex `Python` modules using one or more of these data structures.

WORKING WITH LISTS

Python supports a list data type, along with a rich set of list-related functions. Since lists are not typed, you can create a list of different data types, as well as multidimensional lists. The next several sections show you how to manipulate list structures in Python.

Lists and Basic Operations

A Python list consists of comma-separated values enclosed in a pair of square brackets. The following examples illustrate the syntax for defining a list in Python, and also how to perform various operations on a Python list:

```
>>> list = [1, 2, 3, 4, 5]
>>> list
[1, 2, 3, 4, 5]
>>> list[2]
3
>>> list2 = list + [1, 2, 3, 4, 5]
>>> list2
[1, 2, 3, 4, 5, 1, 2, 3, 4, 5]
>>> list2.append(6)
>>> list2
[1, 2, 3, 4, 5, 1, 2, 3, 4, 5, 6]
>>> len(list)
5
>>> x = ['a', 'b', 'c']
>>> y = [1, 2, 3]
>>> z = [x, y]
>>> z[0]
['a', 'b', 'c']
>>> len(x)
3
```

You can assign multiple variables to a list, provided that the number and type of the variables match the structure. Here is an example:

```
>>> point = [7,8]
>>> x,y = point
>>> x
7
>>> y
8
```

The following example shows you how to assign values to variables from a more complex data structure:

```
>>> line = ['a', 10, 20, (2023,01,31)]
>>> x1,x2,x3,date1 = line
>>> x1
'a'
>>> x2
10
>>> x3
```

```
20
>>> date1
(2023, 1, 31)
```

If you want to access the year/month/date components of the date1 element in the preceding code block, you can do so with the following code block:

```
>>> line = ['a', 10, 20, (2023,01,31)]
>>> x1,x2,x3,(year,month,day) = line
>>> x1
'a'
>>> x2
10
>>> x3
20
>>> year
2023
>>> month
1
>>> day
31
```

If the number and/or structure of the variables do not match the data, an error message is displayed, as shown here:

```
>>> point = (1,2)
>>> x,y,z = point
Traceback (most recent call last):
  File "<stdin>", line 1, in <module>
ValueError: need more than 2 values to unpack
```

If the number of variables that you specify is less than the number of data items, you will see an error message, as shown here:

```
>>> line = ['a', 10, 20, (2014,01,31)]
>>> x1,x2 = line
Traceback (most recent call last):
  File "<stdin>", line 1, in <module>
ValueError: too many values to unpack
```

Lists and Arithmetic Operations

The minimum value of a list of numbers is the first number in the sorted list of numbers. If you reverse the sorted list, the first number is the maximum value. There are several ways to reverse a list, starting with the technique shown in the following code:

```
x = [3,1,2,4]
maxList = x.sort()
minList = x.sort(x.reverse())

min1 = min(x)
max1 = max(x)
```

```
print min1
print max1
```

The output of the preceding code block is here:

```
1
4
```

A second (and better) way to sort a list is shown here:

```
minList = x.sort(reverse=True)
```

A third way to sort a list involves the built-in functional version of the `sort()` method, as shown here:

```
sorted(x, reverse=True)
```

The preceding code snippet is useful when you do not want to modify the original order of the list or you want to compose multiple list operations on a single line.

Lists and Filter-Related Operations

Python enables you to filter a list (also called *list comprehension*) as shown here:

```
mylist = [1, -2, 3, -5, 6, -7, 8]
pos = [n for n in mylist if n > 0]
neg = [n for n in mylist if n < 0]

print pos
print neg
```

You can also specify `if/else` logic in a filter, as shown here:

```
mylist = [1, -2, 3, -5, 6, -7, 8]
negativeList = [n if n < 0 else 0 for n in mylist]
positiveList = [n if n > 0 else 0 for n in mylist]

print positiveList
print negativeList
```

The output of the preceding code block is here:

```
[1, 3, 6, 8]
[-2, -5, -7]
[1, 0, 3, 0, 6, 0, 8]
[0, -2, 0, -5, 0, -7, 0]
```

THE JOIN(), RANGE(), AND SPLIT() FUNCTIONS

Python provides the `join()` method for concatenating text strings, as shown here:

```
>>> parts = ['Is', 'SF', 'In', 'California?']
>>> ' '.join(parts)
```

```
'Is SF In California?'
>>> ','.join(parts)
'Is,SF,In,California?'
>>> ''.join(parts) 'IsSFInCalifornia?'
```

There are several ways to concatenate a set of strings and then print the result. The following is the most inefficient way to do so:

```
print "This" + " is" + " a" + " sentence"
```

Either of the following is preferred:

```
print "%s %s %s %s" % ("This", "is", "a", "sentence")
print " ".join(["This","is","a","sentence"])
```

The next code block illustrates the Python range() function that you can use to iterate through a list, as shown here:

```
>>> for i in range(0,5):
...     print i
...
0
1
2
3
4
```

You can use a for loop to iterate through a list of strings, as shown here:

```
>>> x
['a', 'b', 'c']
>>> for w in x:
...     print w
...
a
b
c
```

You can use a for loop to iterate through a list of strings and provide additional details, as shown here:

```
>>> x
['a', 'b', 'c']
>>> for w in x:
...     print len(w), w
...
1 a
1 b
1 c
```

The preceding output displays the length of each word in the list x, followed by the word itself.

You can use the Python split() function to split the words in a text string and populate a list with those words. An example is here:

```
>>> x = "this is a string"
>>> list = x.split()
>>> list
['this', 'is', 'a', 'string']
```

A simple way to print the list of words in a text string is shown here:

```
>>> x = "this is a string"
>>> for w in x.split():
...     print w
...
this
is
a
string
```

You can search for a word in a string as follows:

```
>>> x = "this is a string"
>>> for w in x.split():
...     if(w == 'this'):
...         print "x contains this"
...
x contains this
...
```

ARRAYS AND THE APPEND() FUNCTION

Although Python does have an array type (import array), which is essentially a heterogeneous list, the array type has no advantages over the list type other than a slight saving in memory use. You can also define heterogeneous arrays:

```
a = [10, 'hello', [5, '77']]
```

You can append a new element to an element inside a list:

```
>>> a = [10, 'hello', [5, '77']]
>>> a[2].append('abc')
>>> a
[10, 'hello', [5, '77', 'abc']]
```

You can assign simple variables to the elements of a list, as shown here:

```
myList = [ 'a', 'b', 91.1, (2014, 01, 31) ]
x1, x2, x3, x4 = myList
print 'x1:',x1
print 'x2:',x2
print 'x3:',x3
print 'x4:',x4
```

The output of the preceding code block is here:

```
x1: a
x2: b
```

```
x3: 91.1
x4: (2014, 1, 31)
```

The `Python split()` function is more convenient (especially when the number of elements is unknown or variable) than the preceding sample, and you will see examples of the `split()` function in the next section.

OTHER LIST-RELATED FUNCTIONS

`Python` provides additional functions that you can use with lists, such as `append()`, `insert()`, `delete()`, `pop()`, and `extend()`. `Python` also supports the functions `index()`, `count()`, `sort()`, and `reverse()`. Examples of these functions are illustrated in the following code block.

Define a `Python` list (notice that duplicates are allowed):

```
>>> a = [1, 2, 3, 2, 4, 2, 5]
```

Display the number of occurrences of 1 and 2:

```
>>> print a.count(1), a.count(2)
1 3
```

Insert −8 in position 3:

```
>>> a.insert(3,-8)
>>> a
[1, 2, 3, -8, 2, 4, 2, 5]
```

Remove occurrences of 3:

```
>>> a.remove(3)
>>> a
[1, 2, -8, 2, 4, 2, 5]
```

Remove occurrences of 1:

```
>>> a.remove(1)
>>> a
[2, -8, 2, 4, 2, 5]
```

Append 19 to the list:

```
>>> a.append(19)
>>> a
[2, -8, 2, 4, 2, 5, 19]
```

Print the index of 19 in the list:

```
>>> a.index(19)
6
```

Reverse the list:

```
>>> a.reverse()
>>> a
[19, 5, 2, 4, 2, -8, 2]
```

Sort the list:

```
>>> a.sort()
>>> a
[-8, 2, 2, 2, 4, 5, 19]
```

Extend list a with list b:

```
>>> b = [100,200,300]
>>> a.extend(b)
>>> a
[-8, 2, 2, 2, 4, 5, 19, 100, 200, 300]
```

Remove the first occurrence of 2:

```
>>> a.pop(2)
2
>>> a
[-8, 2, 2, 4, 5, 19, 100, 200, 300]
```

Remove the last item of the list:

```
>>> a.pop()
300
>>> a
[-8, 2, 2, 4, 5, 19, 100, 200]
```

WORKING WITH LIST COMPREHENSIONS

A list comprehension is a powerful construct in `Python` that enables you to create a list of values in one line of code. Here is a simple example:

```
letters = [w for w in "Chicago Pizza"]
print(letters)
```

If you launch the preceding code snippet you will see the following output:

```
['C', 'h', 'i', 'c', 'a', 'g', 'o', ' ', 'P', 'i', 'z', 'z', 'a']
```

As another example, consider the following two lines of code:

```
names1 = ["Sara","Dave","Jane","Bill","Elly","Dawn"]
names2 = [name for name in names1 if name.startswith("D")]
print("names2:",names2)
```

If you launch the preceding code snippet you will see the following output:

```
names2: ['Dave', 'Dawn']
```

Another example involves a "for ... for ..." construct, as shown here:

```
names3 = ["Sara","Dave"]
names4 = [char for name in names3 for char in name]
```

If you launch the preceding code snippet you will see the following output:

```
names3: ['Sara', 'Dave']
names4: ['S', 'a', 'r', 'a', 'D', 'a', 'v', 'e']
```

The following example illustrates a list comprehension that is an alternative to the map() function:

```
squared = [a*a for a in range(1,10)]
print("squared:",squared)
```

If you launch the preceding code snippet, you will see the following output:

```
squared: [1, 4, 9, 16, 25, 36, 49, 64, 81]
```

The following example illustrates a list comprehension that is an alternative to the filter() function:

```
evens = [a for a in range(1,10) if a%2 == 0]
print("evens:",evens)
```

If you launch the preceding code snippet, you will see the following output:

```
evens: [2, 4, 6, 8]
```

You can also use list comprehensions with two-dimensional arrays, as shown here:

```
import numpy as np
arr1 = np.random.rand(3,3)
maxs = [max(row) for row in arr1]

print("arr1:")
print(arr1)
print("maxs:")
print(maxs)
```

If you launch the preceding code snippet, you will see the following output:

```
arr1:
[[0.8341748  0.16772064 0.79493066]
 [0.876434   0.9884486  0.86085496]
 [0.16727298 0.13095968 0.75362753]]
maxs:
[0.8341747956062362, 0.9884485986312492, 0.7536275263907967]
```

The complete code sample is list_comprehensions.py and is available in the companion files for this chapter.

Now that you understand how to use list comprehensions, the next section shows you how to work with vectors in Python.

WORKING WITH VECTORS

A vector is a one-dimensional array of values, and you can perform vector-based operations, such as addition, subtraction, and inner product. Listing 1.13 displays the contents of MyVectors.py that illustrates how to perform vector-based operations.

LISTING 1.13: MyVectors.py

```
v1 = [1,2,3]
v2 = [1,2,3]
v3 = [5,5,5]

s1 = [0,0,0]
d1 = [0,0,0]
p1 = 0

print("Initial Vectors"
print('v1:',v1)
print('v2:',v2)
print('v3:',v3)

for i in range(len(v1)):
    d1[i] = v3[i] - v2[i]
    s1[i] = v3[i] + v2[i]
    p1    = v3[i] * v2[i] + p1

print("After operations")
print('d1:',d1)
print('s1:',s1)
print('p1:',p1)
```

Listing 1.13 starts with the definition of three lists in Python, each of which represents a vector. The lists d1 and s1 represent the difference of v2 and the sum v2, respectively. The number p1 represents the "inner product" (also called the "dot product") of v3 and v2. The output from Listing 1.13 is here:

```
Initial Vectors
v1: [1, 2, 3]
v2: [1, 2, 3]
v3: [5, 5, 5]
After operations
d1: [4, 3, 2]
s1: [6, 7, 8]
p1: 30
```

WORKING WITH MATRICES

A two-dimensional matrix is a two-dimensional array of values, and you can easily create such a matrix. For example, the following code block illustrates how to access different elements in a 2D matrix:

```
mm = [["a","b","c"],["d","e","f"],["g","h","i"]];
print 'mm:        ',mm
print 'mm[0]:      ',mm[0]
print 'mm[0][1]:',mm[0][1]
```

The output from the preceding code block is here:

```
mm:        ['a', 'b', 'c'], ['d', 'e', 'f'], ['g', 'h', 'i']]
mm[0]:     ['a', 'b', 'c']
mm[0][1]:  b
```

Listing 1.14 displays the contents of My2DMatrix.py that illustrates how to create and populate 2 two-dimensional matrix.

LISTING 1.14: My2DMatrix.py

```
rows = 3
cols = 3

my2DMatrix = [[0 for i in range(rows)] for j in
range(rows)]
print('Before:',my2DMatrix)

for row in range(rows):
  for col in range(cols):
    my2DMatrix[row][col] = row*row+col*col
print('After: ',my2DMatrix)
```

Listing 1.14 initializes the variables rows and cols and then uses them to create the rows x cols matrix my2DMatrix whose values are initially 0. The next part of Listing 1.14 contains a nested loop that initializes the element of my2DMatrix whose position is (row,col) with the value row*row+col*col. The last line of code in Listing 1.14 prints the contents of my2DArray. The output from Listing 1.14 is here:

```
Before: [[0, 0, 0], [0, 0, 0], [0, 0, 0]]
After:  [[0, 1, 4], [1, 2, 5], [4, 5, 8]]
```

QUEUES

A queue is a FIFO ("First In First Out") data structure. Thus, the oldest item in a queue is removed when a new item is added to a queue that is already full.

Earlier in the chapter you learned how to use a Python list to emulate a queue. However, there is also a queue object in Python. The following code snippets illustrate how to use a queue in Python.

```
>>> from collections import deque
>>> q = deque('',maxlen=10)
>>> for i in range(10,20):
...     q.append(i)
```

```
...
>>> print q
deque([10, 11, 12, 13, 14, 15, 16, 17, 18, 19], maxlen=10)
```

The next section shows you how to use tuples in Python.

TUPLES (IMMUTABLE LISTS)

Python supports a data type called a *tuple* that consists of comma-sepa-rated values without brackets (square brackets are for lists, round brackets are for arrays, and curly braces are for dictionaries). Various examples of Python tuples are here:

https://docs.python.org/3.6/tutorial/datastructures.html#tuples-and-sequences

The following code block illustrates how to create a tuple and create new tuples from an existing type in Python.

Define a Python tuple t as follows:

```
>>> t = 1,'a', 2,'hello',3
>>> t
(1, 'a', 2, 'hello', 3)
```

Display the first element of t:

```
>>> t[0]
1
```

Create a tuple v containing 10, 11, and t:

```
>>> v = 10,11,t
>>> v
(10, 11, (1, 'a', 2, 'hello', 3))
```

Try modifying an element of t (which is immutable):

```
>>> t[0] = 1000
Traceback (most recent call last):
  File "<stdin>", line 1, in <module>
TypeError: 'tuple' object does not support item assignment
```

Python "deduplication" is useful because you can remove duplicates from a set and obtain a list, as shown here:

```
>>> lst = list(set(lst))
```

NOTE *The "in" operator on a list to search is O(n) whereas the "in" operator on a set is O(1).*

The next section discusses Python sets.

SETS

A `Python` *set* is an unordered collection that does not contain duplicate elements. Use curly braces or the `set()` function to create sets. Set objects support set-theoretic operations such as union, intersection, and difference.

NOTE `set()` *is required in order to create an empty set because* `{}` *creates an empty dictionary.*

The following code block illustrates how to work with a `Python` set. Create a list of elements:

```
>>> l = ['a', 'b', 'a', 'c']
```

Create a set from the preceding list:

```
>>> s = set(l)
>>> s
set(['a', 'c', 'b'])
```

Test if an element is in the set:

```
>>> 'a' in s
True
>>> 'd' in s
False
>>>
```

Create a set from a string:

```
>>> n = set('abacad')
>>> n
set(['a', 'c', 'b', 'd'])
>>>
```

Subtract n from s:

```
>>> s - n
set([])
```

Subtract s from n:

```
>>> n - s
set(['d'])
>>>
```

The union of s and n:

```
>>> s | n
set(['a', 'c', 'b', 'd'])
```

The intersection of s and n:

```
>>> s & n
set(['a', 'c', 'b'])
```

The exclusive-or of s and n:

```
>>> s ^ n
```

```
set(['d'])
```
The next section shows you how to work with Python dictionaries.

DICTIONARIES

`Python` has a key/value structure called a "dict" that is a hash table. A `Python` dictionary (and hash tables in general) can retrieve the value of a key in constant time, regardless of the number of entries in the dictionary (and the same is true for sets). You can think of a set as essentially just the keys (not the values) of a `dict` implementation.

The contents of a `dict` can be written as a series of `key:value` pairs, as shown here:

```
dict1 = {key1:value1, key2:value2, ... }
```

The "empty dict" is just an empty pair of curly braces `{}`.

Creating a Dictionary

A `Python` dictionary (or hash table) contains of colon-separated key/value bindings inside a pair of curly braces, as shown here:

```
dict1 = {}
dict1 = {'x' : 1, 'y' : 2}
```

The preceding code snippet defines `dict1` as an empty dictionary, and then adds two key/value bindings.

Displaying the Contents of a Dictionary

You can display the contents of `dict1` with the following code:

```
>>> dict1 = {'x':1,'y':2}
>>> dict1
{'y': 2, 'x': 1}
>>> dict1['x']
1
>>> dict1['y']
2
>>> dict1['z']
Traceback (most recent call last):
  File "<stdin>", line 1, in <module>
KeyError: 'z'
```

NOTE *The key/value bindings for a `dict` and a set are not necessarily stored in the same order that you defined them.*

Python dictionaries also provide the get method in order to retrieve key values:

```
>>> dict1.get('x')
1
>>> dict1.get('y')
2
>>> dict1.get('z')
```

As you can see, the Python get method returns None (which is displayed as an empty string) instead of an error when referencing a key that is not defined in a dictionary.

You can also use dict comprehensions to create dictionaries from expressions, as shown here:

```
>>> {x: x**3 for x in (1, 2, 3)}
{1: 1, 2: 8, 3: 37}
```

Checking for Keys in a Dictionary

You can easily check for the presence of a key in a Python dictionary as follows:

```
>>> 'x' in dict1
True
>>> 'z' in dict1
False
```

Use square brackets for finding or setting a value in a dictionary. For example, dict['abc'] finds the value associated with the key 'abc.' You can use strings, numbers, and tuples work as key values, and you can use any type as the value.

If you access a value that is not in the dict, Python throws a KeyError. Consequently, use the "in" operator to check if the key is in the dict. Alternatively, use dict.get(key) which returns the value or None if the key is not present. You can even use the expression get(key, not-found-string) to specify the value to return if a key is not found.

Deleting Keys From a Dictionary

Launch the Python interpreter and enter the following statements:

```
>>> MyDict = {'x' : 5,  'y' : 7}
>>> MyDict['z'] = 13
>>> MyDict
{'y': 7, 'x': 5, 'z': 13}
>>> del MyDict['x']
>>> MyDict
{'y': 7, 'z': 13}
>>> MyDict.keys()
['y', 'z']
```

```
>>> MyDict.values()
[13, 7]
>>> 'z' in MyDict
True
```

Iterating Through a Dictionary

The following code snippet shows you how to iterate through a dictionary:

```
MyDict = {'x' : 5,  'y' : 7, 'z' : 13}

for key, value in MyDict.items():
    print key, value
```

The output from the preceding code block is here:

```
y 7
x 5
z 13
```

Interpolating Data From a Dictionary

The % operator substitutes values from a Python dictionary into a string by name. Listing 1.15 contains an example of doing so.

LISTING 1.15: InterpolateDict1.py

```
hash = {}
hash['beverage'] = 'coffee'
hash['count'] = 3

# %d for int, %s for string
s = 'Today I drank %(count)d cups of %(beverage)s' % hash
print('s:', s)
```

The output from the preceding code block is here:

```
Today I drank 3 cups of coffee
```

DICTIONARY FUNCTIONS AND METHODS

Python provides various functions and methods for a Python dictionary, such as cmp(), len(), and str() that compare two dictionaries, return the length of a dictionary, and display a string representation of a dictionary, respectively.

You can also manipulate the contents of a Python dictionary using the functions clear() to remove all elements, copy() to return a shall copy, get() to retrieve the value of a key, items() to display the (key,value) pairs of a dictionary, keys() to displays the keys of a dictionary, and values() to return the list of values of a dictionary.

The next section discusses other Python sequence types that have not been discussed in previous sections of this chapter.

OTHER SEQUENCE TYPES IN PYTHON

Python supports seven sequence types: str, unicode, list, tuple, bytearray, buffer, and xrange.

You can iterate through a sequence and retrieve the position index and corresponding value at the same time using the enumerate() function.

```
>>> for i, v in enumerate(['x', 'y', 'z']):
...     print i, v
...
0 x
1 y
2 z
```

Bytearray objects are created with the built-in function bytearray(). Although buffer objects are not directly supported by Python syntax, you can create them via the built-in buffer() function.

Objects of type xrange are created with the xrange() function. An xrange object is similar to a buffer in the sense that there is no specific syntax to create them. Moreover, xrange objects do not support operations such as slicing, concatenation or repetition.

At this point you have seen all the Python type that you will encounter in the remaining chapters of this book, so it makes sense to discuss mutable and immutable types in Python, which is the topic of the next section.

MUTABLE AND IMMUTABLE TYPES IN PYTHON

Python represents its data as objects. Some of these objects (such as lists and dictionaries) are mutable, which means you can change their content without changing their identity. Objects such as integers, floats, strings and tuples are objects that cannot be changed. The key point to understand is the difference between changing the value versus assigning a new value to an object; you cannot change a string but you can assign it a different value. This detail can be verified by checking the id value of an object, as shown in Listing 1.16.

LISTING 1.16: Mutability.py

```
s = "abc"
print('id #1:', id(s))
print('first char:', s[0])

try:
  s[0] = "o"
except:
  print('Cannot perform reassignment')

s = "xyz"
print('id #2:',id(s))
s += "uvw"
print('id #3:',id(s))
```

The output of Listing 1.16 is here:

```
id #1: 4297972672
first char: a
Cannot perform reassignment
id #2: 4299809336
id #3: 4299777872
```

Thus, a `Python` type is immutable if its value cannot be changed (even though it's possible to assign a new value to such a type), otherwise a `Python` type is mutable. The `Python` immutable objects are of type `bytes`, `complex`, `float`, `int`, `str`, or `tuple`. However, dictionaries, lists, and sets are mutable. The key in a hash table must be an immutable type.

Since strings are immutable in `Python`, you cannot insert a string in the "middle" of a given text string unless you construct a second string using concatenation. For example, suppose you have the string:

```
"this is a string"
```

and you want to create the following string:

```
"this is a longer string"
```

The following `Python` code block illustrates how to perform this task:

```
text1 = "this is a string"
text2 = text1[0:10] + "longer" + text1[9:]
print 'text1:',text1
print 'text2:',text2
```

The output of the preceding code block is here:

```
text1: this is a string
text2: this is a longer string
```

SUMMARY

This chapter showed you how to work with numbers and perform arithmetic operations on numbers, and then you learned how to work with strings and use string operations. Next, you learned about condition logic, such as `if`/ `elif` statements. You also learned how to work with loops in `Python`, including for loops and while loops.

In addition, you saw how to work with various `Python` data types. In particular, you learned about tuples, sets, and dictionaries. Then you learned how to work with lists and how to use list-related operations to extract sublists.

The next chapter shows you how to work with conditional statements, loops, and user-defined functions in `Python`.

## CHAPTER 2

RECURSION AND COMBINATORICS

This chapter introduces you to recursion that is illustrated in various Python code samples, followed by an introduction to basic concepts in combinatorics, such as combinations and permutations of objects.

The first part of this chapter shows you how to calculate arithmetic series and geometric series using iterative algorithms as well as recursive algorithms. These examples provide a gentler introduction to recursion if you are new to this topic (experienced users will breeze through these code samples). Next you will learn about calculating factorial values of positive integers as well as Fibonacci numbers. Except for the iterative solution to Fibonacci numbers, these code samples do not involve data structures.

The second part of this chapter discusses concepts in combinatorics, such as permutations and combinations. Note that a thorough coverage of combinatorics can fill an entire undergraduate course in mathematics, whereas this chapter contains only some rudimentary concepts.

You might be wondering why recursion is in Chapter 2 instead of a chapter toward the end of the book, and there are several reasons for doing so. First, recursion is indispensable when working with algorithms that solve tasks that involve most data structures, such as singly linked lists, doubly linked lists, queues, stacks, trees, and graphs.

Second, recursive algorithms exist even for a simple data structure such as an array of sorted elements: binary search is such an algorithm, and a recursive as well as iterative solution is discussed in Chapter 3.

Third, some tasks can only be solved via recursive algorithms, such as traversing the elements of a tree or a graph.

If you are new to recursion, be prepared to read the material more than once and also practice working with the code samples, which will lead to a better understanding of recursion. However, you can also skip the material in this chapter until you encounter code samples later in this chapter that involve recursion.

WHAT IS RECURSION?

Recursion-based algorithms can provide very elegant solutions to tasks that would be difficult to implement via iterative algorithms. For some tasks, such as calculating factorial values, the recursive solution and the iterative solution have comparable code complexity.

As a simple example, suppose that we want to add the integers from 1 to n (inclusive), and let n = 10 so that we have a concrete example. If we denote S as the partial sum of successively adding consecutive integers, then we have the following:

```
S = 1
S = S + 2
S = S + 3
 . . .
S = S + 10
```

If we denote S(n) as the sum of the first n positive integers, then we have the following relationship:

```
S(1) = 1
S(n) = S(n-1) + n for n > 1
```

With the preceding observations in mind, the next section contains code samples for calculating the sum of the first n positive integers using an iterative approach and then with recursion.

ARITHMETIC SERIES

This section shows you how to calculate the sum of a set of positive integers, such as the numbers from 1 to n inclusive. The first algorithm uses an iterative approach and the second algorithm uses recursion.

Before delving into the code samples, there is a simple way to calculate the closed form sum of the integers from 1 to n inclusive, which we will denote as S. Then there are two ways to calculate S, as shown here:

```
S = 1 + 2     + 3     + . . . + (n-1) + n
S = n + (n-1) + (n-2) + . . . + 2     + 1
```

There are n columns on the right side of the preceding pair of equations, and each column has the sum equal to (n+1). Therefore the sum of the right side of the equals sign is n*(n+1). Since the left-side of the equals sign has the sum 2*S, we have the following result:

```
2*S = n*(n+1)
```

Now divide both sides by 2 and we get the well-known formula for the arithmetic sum of the first n positive integers:

```
S = n*(n+1)/2
```

Incidentally, the preceding formula was derived by a young student who was bored with performing the calculation manually: that student was Karl F Gauss (in third grade).

Calculating Arithmetic Series (Iterative)

Listing 2.1 displays the contents of the `arith_sum.py` that illustrates how to calculate the sum of the numbers from 1 to n inclusive using an iterative approach.

LISTING 2.1: arith_sum.py

```
def arith_sum(n):
  sum = 0
  for i in range(1,n+1):
    sum += i
  return sum

max = 20
for j in range(2,max+1):
  sum = arith_sum(j)
  print("sum from 1 to",j,"=",sum)
```

Listing 2.1 starts with the function `arith_sum()` that contains a loop that literately adds the numbers from 1 to n. The next portion of Listing 2.1 also contains a loop that iterates through the numbers from 2 to 20 inclusive, and then invokes `arith_sum()` with each value of the loop variable to calculate the sum of the integers from 1 to that value. Launch the code in Listing 2.1 and you will see the following output:

```
sum from 1 to 2 = 3
sum from 1 to 3 = 6
sum from 1 to 4 = 10
sum from 1 to 5 = 15
sum from 1 to 6 = 21
sum from 1 to 7 = 28
sum from 1 to 8 = 36
sum from 1 to 9 = 45
sum from 1 to 10 = 55
sum from 1 to 11 = 66
sum from 1 to 12 = 78
sum from 1 to 13 = 91
sum from 1 to 14 = 105
sum from 1 to 15 = 120
sum from 1 to 16 = 136
sum from 1 to 17 = 153
sum from 1 to 18 = 171
sum from 1 to 19 = 190
sum from 1 to 20 = 210
```

Modify the code in Listing 2.1 to calculate the sum of the squares, cubes, and fourth powers of the numbers from 1 to n, along with your own variations of the code.

Calculating Arithmetic Series (Recursive)

Listing 2.2 displays the contents of the arith_sum_recursive.py that illustrates how to calculate the sum of the numbers from 1 to n inclusive using a recursion.

LISTING 2.2: arith_sum_recursive.py

```
def arith_sum(n):
   if(n == 0):
     return n
   else:
     return n + arith_sum(n-1)

max = 20
for j in range(2,max+1):
   sum = arith_sum(j)
   print("sum from 1 to",j,"=",sum)
```

Listing 2.2 starts with the recursive function arith_sum() that uses conditional logic to return n if n equals the value 0 (which is the terminating case); otherwise the code returns the value of n *plus* the value of arith_sum(n-1). Launch the code in Listing 2.2 and you will see the same output as the previous section.

Calculating Partial Arithmetic Series

Listing 2.3 displays the contents of the arith_partial_sum.py that illustrates how to calculate the sum of the numbers from m to n inclusive, where m and n are two positive integers such that m <= n, using an iterative approach.

LISTING 2.3: arith_partial_sum.py

```
def arith_partial_sum(m,n):
   if(m >= n):
     return 0
   else:
     return n*(n+1)/ - m*(m+1)/2

max = 20
for j in range(2,max+1):
   sum = arith_sum(j)
   print("sum from 1 to",j,"=",sum)
```

Listing 2.3 is straightforward: the function arith_partial_sum() that returns the sum of squares from 1 to n *minus* the sum of squares from 1 to m. This function is invoked in a loop in the second part of Listing 2.3, which calculates the difference of the sum of squares from 2 to 20. Launch the code in Listing 2.3 and you will see the following output:

```
arithmetic sum from 2 to 2 = 2
arithmetic sum from 2 to 3 = 3
```

```
arithmetic sum from 2 to 4 = 7
arithmetic sum from 2 to 5 = 12
arithmetic sum from 2 to 6 = 18
arithmetic sum from 3 to 3 = 3
arithmetic sum from 3 to 4 = 4
arithmetic sum from 3 to 5 = 9
arithmetic sum from 3 to 6 = 15
arithmetic sum from 4 to 4 = 4
arithmetic sum from 4 to 5 = 5
arithmetic sum from 4 to 6 = 11
arithmetic sum from 5 to 5 = 5
arithmetic sum from 5 to 6 = 6
```

Now that you have seen some examples involving arithmetic expressions, next turn to geometric series, which is the topic of the following section.

GEOMETRIC SERIES

This section shows you how to calculate the geometric series of a set of positive integers, such as the numbers from 1 to n inclusive. The first algorithm uses an iterative approach and the second algorithm uses recursion.

Before delving into the code samples, there is a simple way to calculate the closed form sum of the geometric series of integers from 1 to n inclusive, where r is the ratio of consecutive terms in the geometric series. Let S denote the sum, which we can express as follows:

```
S    = 1+ r + r^2 + r^3 + . . . + r^(n-1) + r^n
r*S =     r + r^2 + r^3 + . . . + r^(n-1) + r^n + r^(n+1)
```

Now subtract each term in the second row above from the corresponding term in the first row, and we have the following result:

```
S - r*S = 1 - r^(n+1)
```

Now factor s from both terms on the left side of the preceding equation and we get the following result:

```
S*(1 - r) = 1 - r^(n+1)
```

Now divide both sides of the preceding equation by the term `(1-r)` to get the formula for the sum of the geometric series of the first n positive integers:

```
S = [1 - r^(n+1)]/(1-r)
```

If r = 1 then the preceding equation returns an infinite value, which makes sense because S is the sum of an infinite number of occurrences of the number 1.

Calculating a Geometric Series (Iterative)

Listing 2.4 displays the contents of the geom_sum.py that illustrates how to calculate the sum of the numbers from 1 to n inclusive using an iterative approach.

LISTING 2.4: geom_sum.py

```
def geom_sum(n,ratio):
  partial = 0
  power   = 1
  for i in range(1,n+1):
    partial += power
    power *= ratio
  return partial


max = 10
ratio = 2
for j in range(2,max+1):
  prod = geom_sum(j,ratio)
  print("geometric sum for ratio=",ratio,
"from 1 to",j,"=",prod)
```

Listing 2.4 starts with the function geom_sum() that contains a loop that calculates the sum of the powers of the numbers from 1 to n, where the power is the value of the variable ratio. The second part of Listing 2.4 contains a loop that invokes the function geom_sum() with the values 2, 3, . . ., n, and a fixed value of 2 for the variable ratio. Launch the code in Listing 2.4 and you will see the following output:

```
geometric sum for ratio= 2 from 1 to 2 = 3
geometric sum for ratio= 2 from 1 to 3 = 7
geometric sum for ratio= 2 from 1 to 4 = 15
geometric sum for ratio= 2 from 1 to 5 = 31
geometric sum for ratio= 2 from 1 to 6 = 63
geometric sum for ratio= 2 from 1 to 7 = 127
geometric sum for ratio= 2 from 1 to 8 = 255
geometric sum for ratio= 2 from 1 to 9 = 511
geometric sum for ratio= 2 from 1 to 10 = 1023
```

Calculating Arithmetic Series (Recursive)

Listing 2.5 displays the contents of the geom_sum_recursive.py that illustrates how to calculate the sum of the geometric series of the numbers from 1 to n inclusive using recursion. Note that the following code sample uses *tail recursion*.

LISTING 2.5: geom_sum_recursive.py

```
def geom_sum(n,ratio,term,sum):
  if(n == 1):
    return sum
  else:
    term *= ratio
    sum += term
    return geom_sum(n-1,ratio,term,sum)


max = 10
ratio = 2
```

```
sum = 1
term = 1

for j in range(2,max+1):
  prod = geom_sum(j,ratio,term,sum)
  print("geometric sum for ratio=",ratio,"from 1 to",j,"=",prod)
```

Listing 2.5 contains the function `geom_sum()` that takes four parameters: n (the current value of the upper range), ratio (which is the exponent 2 in this code sample), term (which is the current intermediate term of the sum), and sum (the target sum).

As you can see, the code returns the value 1 when n equals 1; otherwise, the values of term and sum are updated, and the function `geom_sum()` is invoked whose *only* difference is to decrement n by 1.

This code sample illustrates tail recursion, which is more efficient than regular recursion, and perhaps a little more intuitive as well. The second part of Listing 2.5 contains a loop that invokes the function `geom_sum()` as the loop iterates from 2 to max inclusive. Launch the code in Listing 2.5 and you will see the same output as the previous section.

FACTORIAL VALUES

This section contains three code samples for calculating factorial values: one code sample uses a loop and the other two code samples use recursion.

As a reminder, the *factorial* value of a positive integer n is the product of all the numbers from 1 to n (inclusive). Therefore, we have the following values:

```
Factorial(2)   = 2*1 = 2
Factorial(3)   = 3*2*1 = 6
Factorial(4)   = 4*3*2*1 = 24
Factorial(5)   = 5*4*3*2*1 = 120
Factorial(6)   = 6*5*4*3*2*1 = 720
Factorial(7)   = 7*6*5*4*3*2*1 = 5040
```

If you look at the preceding list of calculations, you can see some interesting relationships among factorial numbers:

```
Factorial(3) = 3 * Factorial(2)
Factorial(4) = 4 * Factorial(3)
Factorial(5) = 5 * Factorial(4)
Factorial(6) = 6 * Factorial(5)
Factorial(7) = 7 * Factorial(6)
```

Based on the preceding observations, it's reasonably intuitive to infer the following relationship for factorial numbers:

```
Factorial(1) = 1
Factorial(n) = n * Factorial(n-1) for n > 1
```

The next section uses the preceding formula in order to calculate the factorial value of various numbers.

Calculating Factorial Values (Iterative)

Listing 2.6 displays the contents of the `Factorial1.py` that illustrates how to calculate factorial numbers using an iterative approach.

LISTING 2.6: Factorial1.py

```
def factorial(n):
    prod = 1
    for i in range(1,n+1):
        prod *= i
    return prod

max = 20
for n in range(0,max):
    result = factorial(n)
    print("factorial",n,"=",result)
```

Listing 2.6 starts with the function `factorial()` that contains a loop to multiply the numbers from 1 to n and storing the product in the variable prod whose initial value is 1. The second part of Listing 2.6 contains a loop that invokes `factorial()` with the loop variable that ranges from 0 to max. Launch the code in Listing 2.6 and you will see the following output:

```
factorial 0 = 1
factorial 1 = 1
factorial 2 = 2
factorial 3 = 6
factorial 4 = 24
factorial 5 = 120
factorial 6 = 720
factorial 7 = 5040
factorial 8 = 40320
factorial 9 = 362880
factorial 10 = 3628800
factorial 11 = 39916800
factorial 12 = 479001600
factorial 13 = 6227020800
factorial 14 = 87178291200
factorial 15 = 1307674368000
factorial 16 = 20922789888000
factorial 17 = 355687428096000
factorial 18 = 6402373705728000
factorial 19 = 121645100408832000
```

Calculating Factorial Values (Recursive)

Listing 2.7 displays the contents of the `Factorial2.py` that illustrates how to calculate factorial values using recursion.

LISTING 2.7: Factorial2.py

```
def factorial(n):
    if(n <= 1):
```

```
      return 1
   else:
      return n * factorial(n-1)

max = 20
for n in range(0,max):
   result = factorial(n)
   print("factorial",n,"=",result)
```

Listing 2.7 starts with the function `factorial()` that is the same function that you saw in Listing 2.6. Notice that the second portion of Listing 2.7 is the same as the second portion of Listing 2.6. Now launch the code in Listing 2.7 and you will see the same output as the preceding example.

Calculating Factorial Values (Tail Recursion)

Listing 2.8 displays the contents of the `Factorial3.py` that illustrates how to calculate factorial values using tail recursion.

LISTING 2.8: Factorial3.py

```
def factorial(n, prod):
   if(n <= 1):
      return prod
   else:
      return factorial(n-1, n*prod)

max = 20
for n in range(0,max):
   result = factorial(n, 1)
   print("factorial",n,"=",result)
```

Listing 2.8 starts with the recursive function `factorial()` that uses tail recursion, which is somewhat analogous to the tail recursion in Listing 2.5. The second portion of Listing 2.8 is the same as the second portion of Listing 2.5. Launch the code in Listing 2.8 and you will see the same output as the preceding example.

FIBONACCI NUMBERS

Fibonacci numbers are simple yet interesting, and also appear in nature (such as the pattern of sunflower seeds). Here is the definition of the `Fibonacci` sequence:

```
Fib(0) = 0
Fib(1) = 1
Fib(n) = Fib(n-1)+Fib(n-2) for n >= 2
```

Note that it's possible to specify different "seed" values for `Fib(0)` and `Fib(1)`, but the values 0 and 1 are the most commonly used values.

Calculating Fibonacci Numbers (Recursive)

Listing 2.9 displays the contents of the Fibonacci1.py that illustrates how to calculate Fibonacci numbers using recursion.

LISTING 2.9: Fibonacci1.py

```
# very inefficient:
def fibonacci(n):
  if n <= 1:
    return n
  else:
    return fibonacci(n-2) + fibonacci(n-1)

max=20
for i in range(0,max):
  fib = fibonacci(i)
  print("fibonacci",i,"=",fib)
```

Listing 2.9 starts the recursive function fibonacci() that returns 1 if n equals 1. If n is greater than 1, the code returns the sum of *two* invocations of fibonacci(): the first with the value n-2 and the second with the value n-1.

The second part of Listing 2.9 contains another loop that invokes the function fibonacci() with the values of the loop variable that iterates from 0 to max. Now launch the code in Listing 2.9 and you will see the following output:

```
fibonacci 0 = 0
fibonacci 1 = 1
fibonacci 2 = 1
fibonacci 3 = 2
fibonacci 4 = 3
fibonacci 5 = 5
fibonacci 6 = 8
fibonacci 7 = 13
fibonacci 8 = 21
fibonacci 9 = 34
fibonacci 10 = 55
fibonacci 11 = 89
fibonacci 12 = 144
fibonacci 13 = 233
fibonacci 14 = 377
fibonacci 15 = 610
fibonacci 16 = 987
fibonacci 17 = 1597
fibonacci 18 = 2584
fibonacci 19 = 4181
```

Calculating Fibonacci Numbers (Iterative)

Listing 2.10 displays the contents of the Fibonacci2.py that illustrates how to calculate Fibonacci numbers using an iterative approach.

LISTING 2.10: Fibonacci2.py

```
import numpy as np

max=20
arr1 = np.zeros(max)
arr1[0] = 0
arr1= 1

for i in range(2,max):
    arr1[i] = arr1[i-1] + arr1[i-2]
    print("fibonacci",i,"=",arr1[i])
```

Listing 2.10 also calculates the values of Fibonacci numbers; however, this code sample stores intermediate values in an array. Despite the overhead of an array, this code is much more efficient than the code in Listing 2.9. Now launch the code in Listing 2.10 and you will see the same output as the previous section.

TASK: REVERSE A STRING VIA RECURSION

Listing 2.11 displays the contents of the Python file reverse.py that illustrates how to use recursion in order to reverse a string.

LISTING 2.11: reverse.py

```
import numpy as np

def reverser(str):
    if(str == None or len(str) == 0):
        return str
    print("all-but-first chars:",str[1:])
    return reverser(str[1:])+list(str[0])

names = np.array(["Nancy", "Dave", "Dominic"])

for name in names:
    str_list = list(name)
    result = reverser(str_list)
    print("=> Word: ",name," reverse: ",result)
    print()
```

Listing 2.11 starts with the recursive function reverser() that invokes itself with a substring that omits the first character, which is appended to the result of invoking reverser() recursively, as shown here:

```
return reverser(str[1:])+list(str[0])
```

The second part of Listing 2.11 contains a loop that invokes the reverser() method with different strings that belong to an array. Launch the code in Listing 2.11 and you will see the following output:

```
all-but-first chars: ['a', 'n', 'c', 'y']
all-but-first chars: ['n', 'c', 'y']
all-but-first chars: ['c', 'y']
all-but-first chars: ['y']
all-but-first chars: []
=> Word:  Nancy  reverse:  ['y', 'c', 'n', 'a', 'N']


all-but-first chars: ['a', 'v', 'e']
all-but-first chars: ['v', 'e']
all-but-first chars: ['e']
all-but-first chars: []
=> Word:  Dave  reverse:  ['e', 'v', 'a', 'D']


all-but-first chars: ['o', 'm', 'i', 'n', 'i', 'c']
all-but-first chars: ['m', 'i', 'n', 'i', 'c']
all-but-first chars: ['i', 'n', 'i', 'c']
all-but-first chars: ['n', 'i', 'c']
all-but-first chars: ['i', 'c']
all-but-first chars: ['c']
all-but-first chars: []
=> Word:  Dominic  reverse:  ['c', 'i', 'n', 'i', 'm', 'o', 'D']
```

TASK: CHECK FOR BALANCED PARENTHESES

This task involves only round parentheses: later you will see an example of checking for balanced parentheses that can include square brackets and curly braces. Here are some examples of strings that contain round parentheses:

```
S1 = "()()()"
S2 = "(()()())"
S3 = "()("
S4 = "(())"
S5 = "()()("
```

As you can see, the strings S1, S2, and S4 have balanced parentheses, whereas the strings S2 and S5 has unbalanced parentheses.

Listing 2.12 displays the contents of the Python file balanced_parentheses.py that illustrates how to determine whether or not a string consists of balanced parentheses.

LISTING 2.12: balanced_parentheses.py

```
import numpy as np

def check_balanced(text):
  counter = 0
  text_len = len(text)

  for i in range(text_len):
    if (text[i] == '('):
```

```
      counter += 1
    else:
      if (text[i] == ')'):
        counter -= 1

    if (counter < 0):
      break

  if (counter == 0):
    print("balanced string:",text)
  else:
    print("unbalanced string:",text)
  print()

exprs = np.array(["()()()", "(()()())","()(","(())","()()
("])

for str in exprs:
  check_balanced(str)
```

Listing 2.12 starts with the iterative function `check_balanced()` that uses conditional logic to check the contents of the current character in the input string. The code increments the variable `counter` if the current character is a left parenthesis "(", and decrements the variable `counter` if the current character is a right parentheses ")". The only way for an expression to consist of a balanced set of parentheses is for counter to equal 0 when the loop has finished execution.

The second part of Listing 2.12 contains a loop that invokes the function `check_balanced()` with different strings that are part of an array of strings. Launch the code in Listing 2.12 and you will see the following output:

```
exprs = np.array(["()()()", "(()()())","()(","(())","()()("])
balanced string: ()()()

balanced string: (()()())

unbalanced string: ()(

balanced string: (())

unbalanced string: ()()(
```

TASK: CALCULATE THE NUMBER OF DIGITS

Listing 2.13 displays the contents of the Python file `count_digits.py` that illustrates how to calculate the number of digits in positive integers.

LISTING 2.13: count_digits.py

```
import numpy as np

def count_digits(num, result):
  if( num == 0 ):
    return result
  else:
    #print("new result:",result+1)
    #print("new number:",int(num/10))
    return count_digits(int(num/10), result+1)

numbers = np.array([1234, 767, 1234321, 101])

for num in numbers:
  result = count_digits(num, 0)
  print("Digits in ",num," = ",result)
```

Listing 2.13 starts with the Python function count_digits() that recursively invokes itself with the term int(num/10), where num is the input parameter. Moreover, each invocation of count_digits() increments the value of the parameter result. Eventually num will be equal to 0 (the terminating condition), at which point the value of result is returned.

If the logic of this code is not clear to you, try tracing through the code with the numbers 5, 25, 150, and you will see that the function count_digits() returns the values 1, 2, and 3, respectively. Now launch the code in Listing 2.13 and you will see the following output:

```
Digits in   1234    =   4
Digits in   767   =   3
Digits in   1234321   =   7
Digits in   101   =   3
```

TASK: DETERMINE IF A POSITIVE INTEGER IS PRIME

Listing 2.14 displays the contents of the Python file check_prime.py that illustrates how to calculate the number of digits in positive integers.

LISTING 2.14: check_prime.py

```
import numpy as np

PRIME = 1
COMPOSITE = 0

def is_prime(num):
  div = 2

  while(div*div < num):
   if( num % div != 0):
      div += 1
```

```
    else:
        return COMPOSITE
    return PRIME


upperBound = 20

for num in range(2, upperBound):
    result = is_prime(num)
    if(result == True):
        print(num,": is prime")
    else:
        print(num,": is not prime")
```

Listing 2.14 starts with the Python function is_prime() that contains a loop that checks whether or not any integer in the range of 2 to sqrt(num) divides the parameter num, and then returns the appropriate result.

The second portion of Listing 2.14 contains a loop that iterates through the numbers from 2 to upperBound (which has the value 20) to determine which numbers are prime. Now launch the code in Listing 2.14 and you will see the following output:

```
2 : is prime
3 : is prime
4 : is not prime
5 : is prime
6 : is not prime
7 : is prime
8 : is not prime
9 : is not prime
10 : is not prime
11 : is prime
12 : is not prime
13 : is prime
14 : is not prime
15 : is not prime
16 : is not prime
17 : is prime
18 : is not prime
19 : is prime
```

TASK: FIND THE PRIME FACTORIZATION OF A POSITIVE INTEGER

Listing 2.15 displays the contents of the Python file prime_divisors.py that illustrates how to find the prime divisors of a positive integer.

LISTING 2.15: prime_divisors.py

```
import numpy as np

PRIME = 1
COMPOSITE = 0
```

```python
def is_prime(num):
  div = 2

  while(div < num):
    if( num % div != 0):
      div += 1
    else:
      return COMPOSITE

  #print("found prime:",num)
  return PRIME

def find_prime_divisors(num):
  div = 2
  prime_divisors = ""

  while(div <= num):
    prime = is_prime(div)

    if(prime == True):
      #print("=> prime number:",div)
      if(num % div == 0):
        prime_divisors += " "+str(div)
        num = int(num/div)
      else:
        div += 1
    else:
      div += 1

  return prime_divisors

upperBound = 20

for num in range(4, upperBound):
  result = find_prime_divisors(num)
  print("Prime divisors of ",num,":",result)
```

Listing 2.15 starts with the Python function `is_prime()` from Listing 2.14 that determines whether or not a positive integer is a prime number. Next, the Python function `find_prime_divisors()` contains a loop that iterates through the integers from 2 to num that checks which of those numbers is a prime number.

When a prime number is found, the code checks if that prime number is also a divisor of num: if so, that prime divisor is appended to the string `prime_divisors`. The final portion of Listing 2.15 returns the string `prime_divisors` that contains the prime factorization of the parameter num. Now launch the code in Listing 2.15 and you will see the following output:

```
Prime divisors of  2 :   2
Prime divisors of  4 :   2 2
Prime divisors of  5 :   5
Prime divisors of  6 :   2 3
```

```
Prime divisors of  7 :  7
Prime divisors of  8 :  2 2 2
Prime divisors of  9 :  3 3
Prime divisors of  10 :  2 5
Prime divisors of  11 :  11
Prime divisors of  12 :  2 2 3
Prime divisors of  13 :  13
Prime divisors of  14 :  2 7
Prime divisors of  15 :  3 5
Prime divisors of  16 :  2 2 2 2
Prime divisors of  17 :  17
Prime divisors of  18 :  2 3 3
Prime divisors of  19 :  19
```

TASK: GOLDBACH'S CONJECTURE

Goldbach's conjecture states that every even number greater than 3 can be expressed as the sum of two odd prime numbers.

Listing 2.16 displays the contents of the Python file goldbach_conjecture.py that illustrates how to determine a pair of prime numbers whose sum equals a given even number.

LISTING 2.16: goldbach_conjecture.py

```python
import numpy as np

PRIME = 1
COMPOSITE = 0

def prime(num):
  div = 2

  while(div < num):
   if( num % div != 0):
      div += 1
   else:
      return COMPOSITE
   return PRIME

def find_prime_factors(even_num):
  for num in range(3, int(even_num/2)):
    if(prime(num) == 1):
      if(prime(even_num-num) == 1):
        print(even_num , " = " , num , "+" , (even_num-
num))

upperBound = 30

for num in range(4, upperBound):
  find_prime_factors(num)
```

Listing 2.16 also starts with the function `prime()` that determines whether or not the parameter `num` is a prime number. Next, the function `find_prime_factors()` contains a loop whose loop variable `num` iterates from 3 to half the value of the parameter `even_num`. If `num` is a prime number, then the conditional logic in Listing 2.16 invokes `prime()` with the number `even_num-num`.

If both `num` and `even_num` are prime, then they are a solution to Goldbach's conjecture because the sum of these two numbers equals the parameter `even_num`. Now launch the code in Listing 2.16 and you will see the following output:

```
8   =   3 + 5
10  =   3 + 7
12  =   5 + 7
14  =   3 + 11
16  =   3 + 13
16  =   5 + 11
18  =   5 + 13
18  =   7 + 11
20  =   3 + 17
20  =   7 + 13
22  =   3 + 19
22  =   5 + 17
24  =   5 + 19
24  =   7 + 17
24  =   11 + 13
26  =   3 + 23
26  =   7 + 19
28  =   5 + 23
28  =   11 + 17
```

As you can see from the preceding output, the numbers 16, 18, 20, 22, 26, and 28 have two solutions to Goldbach's conjecture, and the number 24 has three such solutions.

TASK: CALCULATE THE GCD (GREATEST COMMON DIVISOR)

Listing 2.17 displays the contents of the Python file `gcd.py`, which is the first of two solutions for calculating the GCD of two positive integers (both solutions rely on Euclid's algorithm).

LISTING 2.17: gcd.py

```python
import numpy as np

def gcd(num1, num2):
  if(num1 % num2 == 0):
   return num2;
  elif (num1 < num2):
   #print("Switching",num1,"and",num2)
   return gcd(num2, num1);
  else:
   #print("Reducing",num1,"and",num2)
   return gcd(num1-num2, num2)
```

```
arr1 = np.array([24, 36, 50, 100, 200])
arr2 = np.array([10, 18, 11,  64, 120])

for i in range(0,len(arr1)):
  num1 = arr1[i]
  num2 = arr2[i]
  result = gcd(num1,num2)
  print("The GCD of",num1,"and",num2,"=",result)
```

Listing 2.17 starts with the `Python` function `gcd()` that takes two parameters and repeatedly subtracts the smaller from the larger, and simultaneously invoking itself recursively. Eventually `num1` % `num2` equals zero, at which point the GCD equals `num2`, which is the value that is returned.

The second portion of Listing 2.17 contains a loop that iterates through the values of two arrays of positive integers; during each iteration, the function `gcd()` is invoked with a pair of corresponding numbers from the two arrays. Now launch the code in Listing 2.17 and you will see the following output:

```
The GCD of 24 and 10 = 2
The GCD of 36 and 18 = 18
The GCD of 50 and 11 = 1
The GCD of 100 and 64 = 4
The GCD of 200 and 120 = 40
```

Listing 2.18 displays the contents of `simple_gcd.py` that is a more concise way to compute the GCD of two positive integers (and also uses recursion).

LISTING 2.18: simple_gcd.py

```
import numpy as np

def gcd(x1, x2):
  if not x2:
    return x1
  return gcd(x2, x1 % x2)

arr1 = np.array([10, 24, 50, 17, 100])
arr2 = np.array([24, 10, 15, 17, 1250])

for idx in range(0,len(arr1)):
  num1 = arr1[idx]
  num2 = arr2[idx]
  result = gcd(num1,num2)
  print("gcd of",num1,"and",num2,"=",result)
```

Listing 2.19 is a more compact implementation of Euclid's algorithm that achieves the same result as Listing 2.18; if the logic is unclear, review the details of Listing 2.18 to convince yourself that the logic in both code samples is the same. Now launch the code in Listing 2.19 and you will see the following output:

```
gcd of 10 and 24 = 2
gcd of 24 and 10 = 2
gcd of 50 and 15 = 5
gcd of 17 and 17 = 17
gcd of 100 and 1250 = 50
```

Now that we can calculate the GCD of two positive integers, we can use this code to easily calculate the LCM (lowest common multiple) of two positive integers, as discussed in the next section.

TASK: CALCULATE THE LCM (LOWEST COMMON MULTIPLE)

Listing 2.19 displays the contents of the Python file simple_lcm.py that illustrates how to calculate the LCM of two positive integers.

LISTING 2.19: simple_lcm.py

```python
import numpy as np

def gcd(x1, x2):
  if not x2:
    return x1
  return gcd(x2, x1 % x2)

def lcm(num1, num2):
  gcd1 = gcd(num1, num2)
  lcm1 = num1*num2/gcd1

  return lcm1

arr1 = np.array([24, 36, 50, 100, 200])
arr2 = np.array([10, 18, 11,  64, 120])

for i in range(0,len(arr1)):
  num1 = arr1[i]
  num2 = arr2[i]
  result = lcm(num1,num2)
  print("The LCM of",num1,"and",num2,"=",result)
```

Listing 2.19 contains the function gcd() to calculate the GCD of two positive integers. The next function lcm() calculates the LCM of two numbers num1 and num2 by making the following observation:

```
LCM(num1, num2) = num1*num2/GCD(num1,num2)
```

The final portion of Listing 2.19 contains a loop that iterates through two arrays of positive integers to calculate the LCM of pairs of integers. Now launch the code in Listing 2.19 and you will see the following output:

```
The LCM of 24 and 10 = 120.0
The LCM of 36 and 18 = 36.0
```

```
The LCM of 50 and 11 = 550.0
The LCM of 100 and 64 = 1600.0
The LCM of 200 and 120 = 600.0
```

This concludes the portion of the chapter regarding recursion. The next section introduces you to combinatorics (a well-known branch of mathematics), along with some code samples for calculating combinatorial values and the number of permutations of objects.

WHAT IS COMBINATORICS?

In simple terms, combinatorics involves finding formulas for counting the number of objects in a set. For example, how many different ways can five books can be ordered (i.e., displayed) on a book shelf? The answer involves permutations, which in turn is a factorial value; in this case, the answer is 5! = 120.

As a second example, suppose how many different ways can you select three books from a shelf that contains five books? The answer to this question involves combinations. Keep in mind that if you select three books labeled A, B, and C, then any permutation of these three books is considered the same (the set {A, B, C} and the set {B, A, C} are considered the same selection).

As a third example, how many 5-digit binary numbers contain exactly three 1 values? The answer to this question also involves calculating a combinatorial value. In case you're wondering, the answer is C(5,3) = 5!/[3! * 2!] = 10, provided that we allow for leading zeroes. In fact, this is also the answer to the preceding question about selecting different subsets of books.

You can generalize the previous question by asking how many 4-digit, 5-digit, and 6-digit numbers contain exactly three 1s? The answer is the sum of these values (provided that leading zeroes are permitted):

```
C(4,3) + C(5,3) + C(6,3) = 4 + 10 + 20 = 34
```

Working With Permutations

Consider the following task: given six books, how many ways can you display them side by side? The possibilities are listed here:

position #1: 6 choices
position #2: 5 choices
position #3: 4 choices
position #4: 3 choices
position #5: 2 choices
position #6: 1 choices

The answer is 6x5x4x3x2x1 = 6! = 720. In general, if you have n books, there are n! different ways that you can order them (i.e., display them side by side).

Working With Combinations

Here is a slightly different question: how many ways can you select three books from those six books? Here's the first approximation:

- position #1: six choices
- position #2: five choices
- position #3: four choices

Since the number of books in any position is independent of the other positions, the first answer *might* be 6x5x4 = 120. However, this answer is incorrect because it includes different orderings of three books, but the sequence of books (A, B, C) is the same as (B, A, C) and every other re-ordering of the letters A, B, and C.

As a concrete example, suppose that the books are labeled book #1, book #2, . . . , book #6, and suppose that you select book #1, book #2, and book #3. Here list a list of all the different orderings of those three books:

```
123
132
213
231
312
321
```

The number of different permutations of three books is 3x2x1 = 3! = 6. However, from the standpoint of purely selecting three books, we must treat all six orderings as the same. Therefore, the preceding list of six orderings are indistinguishable from each other. As a result, we must divide the number of permutations by the number of orderings that are considered the same. As a result, the correct answer is N = 6x5x4/[3x2x1] = 120/3! = 120/6 = 20.

Now watch what happens when we multiply the numerator and the denominator of the number N by 3x2x1:

```
N = 6x5x4/[3x2x1] = 6x5x4x3x2x1/[3x2x1 x 3x2x1] = 6!/[3! x 3!]
```

If we perform the preceding task of selecting three books from eight books instead of six books, we get this result:

```
8x7x6/[3x2x1] = 8x7x6x5x4x3x2x1/[3x2x1 * 5x4x3x2x1] = 8!/[3! * 5!]
```

Now suppose you select twelve books from a set of thirty books. The number of ways that this can be done is shown here:

```
30x29x28x...x19/[12x11x...x2x1]
= 30x29x28x...x19x18x17x16x...x2x1/[12x11x...x2x1 * 18x17x16x...x2x1]
= 30!/[12! * 18!]
```

The general formula for calculating the number of ways to select k books from n books is n!/[k! * (n-k)!], which is denoted by the term C(n, k).

Incidentally, if we replace `k` by `n-k` in the preceding formula we get this result:

```
n!/[(n-k)! * (n-(n-k))!] = n!/[(n-k)! * k)!] = C(n,k)
```

Notice that the left-side of the preceding snippet equals `C(n,n-k)`, and therefore we have shown that `C(n,n-k) = C(n,k)`

TASK: CALCULATE THE SUM OF BINOMIAL COEFFICIENTS

Recall from the previous section that the value of the binomial coefficient `C(n,k)` can be computed as follows:

```
C(n,k) = n!/[k! * (n-k)!]
```

Given any positive integer n, the following result (details are in the next section) is true:

```
2**n = C(n,0)+C(n,1)+C(n,2)+. . . +C(n,n-1)+C(n,n)
```

Listing 2.20 displays the contents of the Python file `sum_binomial.py` that calculates the sum of a set of binomial coefficients.

LISTING 2.20: sum_binomial.py

```python
import numpy as np

def factorial(num):
  fact = 1
  for i in range(0,num):
    fact *= (i+1)
  return int(fact)

def binom_coefficient(n,k):
  global fact_values
  coeff = fact_values[n]/[fact_values[k] * fact_values[n-k]]
  #print("calculated coeff:",coeff)
  return int(coeff)

def sum_binomials(exp):
  binomials = np.array([]).astype(int)
  coeff_sum = 0
  for num in range(0,exp+1):
    coeff_value = binom_coefficient(exp,num)
    #print("n:",exp-2,"k:",num,"found coeff_value:",coeff_value)
    coeff_sum += coeff_value

  print("sum of binomial coefficients for",exp,"=",int(coeff_sum))
```

```
exponent = 12
# populate an array with factorial values:
fact_values = np.array([]).astype(int)
for j in range(0,exponent):
  fact = factorial(j)
  fact_values = np.append(fact_values,fact)

for exp in range(1,exponent-1):
  sum_binomials(exp)
```

Listing 2.20 starts with the function `factorial()` to calculate the factorial value of a positive integer (whose code you saw earlier in this chapter). Next, the `Python` function `binom_coefficient()` calculates the binomial of two integers whose formula was derived in a previous section.

The third function is `sum_binomials()` that calculate the sum of a range of binomial values by invoking the function `binom_coefficient()`, where the latter invokes the function `factorial()`. Now launch the code in Listing 2.20 and you will see the following output:

```
sum of binomial coefficients for 1 = 2
sum of binomial coefficients for 2 = 4
sum of binomial coefficients for 3 = 8
sum of binomial coefficients for 4 = 16
sum of binomial coefficients for 5 = 32
sum of binomial coefficients for 6 = 64
sum of binomial coefficients for 7 = 128
sum of binomial coefficients for 8 = 256
sum of binomial coefficients for 9 = 512
sum of binomial coefficients for 10 = 1024
```

THE NUMBER OF SUBSETS OF A FINITE SET

In the preceding section, if we allow `k` to vary from 0 to n inclusive, then we are effectively looking at all possible subsets of a set of n elements, and the number of such sets equals 2^n. We can derive the preceding result in two ways.

Solution #1

The first way is the shortest explanation (and might seem like clever hand waving) and it involves visualizing a row of n books. In order to find every possible subset of those n books, we need only consider that there are two actions for the first position: either the book is selected or it is not selected.

Similarly, there are two actions for the second position: either the second book is selected or it is not selected. In fact, for every book in the set of n books there are the same two choices. Keeping in mind that the selection (or not) of a book in a given position is independent of the selection of the books in every other position, the number of possible choices equals $2 \times 2 \times ... \times 2$ (n times) = 2^n.

Solution #2

Recall the following formulas from algebra:

```
(x+y)^2 = x^2 + 2*x*y + y^2
        = C(2,0)*x^2 + C(2,1)*x*y + C(2,2)*y^2

(x+y)^3 = x^3 + 4*x^2*y + 6*x*x*y^2 + 4*x*y^2 + y^3
        = C(3,0)*x^3 + C(3,0)*x^2*y + C(3,0)*x^x*y^2 +
C(3,0)*x*y^2 + C(3,0)*y^3
```

In general, we have the following formula:

```
              n
(x+y)^n =   SUM C(n,k)*x^k*y^(n-k)
            k=0
```

Now set `x=y=1` in the preceding formula and we get the following result:

```
          n
2^n =   SUM C(n,k)
        k=0
```

The right side of the preceding formula is the sum of the number of all possible subsets of a set of n elements, which the left side shows is equal to `2^n`.

TASK: SUBSETS CONTAINING A VALUE LARGER THAN K

The more complete description of the task for this section is as follows: given a set N of numbers and a number k, find the number of subsets of N that contain at least one number that is larger than k. This *counting* task is an example of a coding task that can easily be solved as a combinatorial problem: you might be surprised to discover that the solution involves a single (and simple) line of code. Define the following set of variables:

```
• N       = a set of numbers
• |N|     = # of elements in N (= n)
• NS      = the non-empty subsets of N
• P(NS)   = the number of non-empty subsets of N ( = |NS|)
• M       = the numbers {n| n < k} where n is an element of N
• |M|     = # of elements in M (= m)
• MS      = the non-empty subsets of M
• P(MS)   = the number of non-empty subsets of M (= |MS|)
• Q       = subsets of N that contain at least one number
            larger than k
```

Note that the set NS is partitioned into the sets Q and M, and that the union of Q and M is NS. In other words, a nonempty subset of N is either in Q or in M, but not in both. Therefore, the solution to the task can be expressed as: `|Q| =` `P(NS) - P(MS)`.

Moreover, the sets in M do not contain any number that is larger than k, which means that no element (i.e., subset) in M is an element of Q, and conversely, no element of Q is an element of M.

Recall from a previous result in this chapter that if a set contains m elements, then the number of subsets is 2**m, and the number of *nonempty* subsets is 2**m − 1. Therefore, the answer for this task is (2**n − 1) − (2**m − 1).

Listing 2.21 displays the contents of subarrays_max_k.py that calculates the sum of a set of binomial coefficients.

LISTING 2.21: subarrays_max_k.py

```
import numpy as np
# Time Complexity: O(1)


#################################################
# N = a set with n elements
# M = a set with m elements
#
# N has 2^n - 1 non-empty subsets
# M has 2^m - 1 non-empty subsets
#
# O = subsets of N with at least one element > k
# P = subsets of N with all numbers <= k
#
# |P| = 2**m-1
# and |O| = |N| - |P| = (2**n-1) - (2**m-1)
#################################################

# number of subarrays whose maximum element > k
def count_subsets(n, m):
  count = (2**n - 1) - (2**m - 1)
  return count

arr = [1, 2, 3, 4, 5]
print("contents of array: ")
for num in arr:
  print(num,end=" ")
  print()

arrk = [1,2,3,4]
for overk in arrk:
  arr_len = len(arr)
  count = count_subsets(arr_len, overk)

  print("overk:    ",overk)
  print("count:    ",count)
  print("--------------")
```

Listing 2.21 contains the Python code that implements the details that are described at the beginning of this section. Although the set N in Listing 2.21 contains a set of consecutive integers from 1 to n, the code works correctly for unsorted arrays or arrays that do not contain consecutive integers. In the latter case, you would need a code block to count the number of elements that are less than a given value of k.

SUMMARY

This chapter started with an introduction to recursion, along with various code samples that involve recursion, such as calculating factorial values, Fibonacci numbers, the sum of an arithmetic series, the sum of a geometric series, the GCD of a pair of positive integers, and the LCM of a pair of positive integers.

Finally, you learned about concepts in combinatorics, and how to derive the formula for the number of permutations and the number of combinations of sets of objects.

## CHAPTER 3

STRINGS AND ARRAYS

This chapter contains Python-based code samples that solving various tasks involving strings and arrays. The code samples in this chapter consists of the following sequence: examples that involve scalars and strings, followed by examples involving vectors (explained further at the end of this introduction), and then some examples involving 2D matrices. In addition, the first half of Chapter 2 is relevant for the code samples in this chapter that involve recursion.

The first part of this chapter starts with a quick overview of the time complexity of algorithms, followed by various Python code samples such as finding palindromes, reversing strings, and determining if the characters in a string are unique.

The second part of this chapter discusses 2D arrays, along with NumPy-based code samples that illustrate various operations that can be performed on 2D matrices. This section also discusses 2D matrices, which are 2D arrays, along with some tasks that you can perform on them. This section also discusses multidimensional arrays, which have properties that are analogous to lower-dimensional arrays.

One other detail to keep in mind pertains to the terms vectors and arrays. In mathematics, a vector is a one-dimensional construct, whereas an array has at least two dimensions. In software development, an array can refer to a one-dimensional array or a higher-dimensional array (depending on the speaker). In this book a vector is always a one-dimensional construct. However, the term array always refers to a one-dimensional array; higher dimensional arrays will be referenced as "2D array," "3D array," and so forth. Therefore, the tasks involving 2D arrays start from the section titled "Working With 2D Arrays".

TIME AND SPACE COMPLEXITY

Algorithms are assessed in terms of the amount of space (based on input size) and the amount of time required for the algorithms to complete their execution, which is represented by "big O" notation. There are three types of time complexity: best case, average case, and worst case. Keep in mind that an algorithm with very good best case performance can have a relatively poor worse case performance.

Recall that O(n) means that an algorithm executes in linear time because its complexity is bounded above and below by a linear function. For example, if three algorithms require 2*n, 5*n, or n/2 operations, respectively, then all of them have O(n) complexity.

Moreover, if the best, average, and worst time performance for a linear search is 1, n/2, and n operations, respectively, then those operations have O(1), O(n), and O(n), respectively. In general, if there are two solutions T1 and T2 for a given task such T2 is more efficient than T1, then T2 requires either less time or less memory. For example, if T1 is an iterative solution for calculating factorial values (or Fibonacci numbers) and T2 involves a recursive solution, then T1 is more efficient than T2 in terms of time, but T1 also requires an extra array to store intermediate values.

The *time-space trade-off* refers to reducing either the amount of time or the amount of memory that is required for executing an algorithm, which involves choosing one of the following:

- execute in less time and more memory
- execute in more time and less memory

Although reducing both time and memory is desirable, it's also a more challenging task. Another point to keep in mind is the following inequalities (logarithms can be in any base that is greater than or equal to 2) for any positive integer $n > 1$:

```
O(log n) < O(n) < O(n*log n) < O(n^2)
```

In addition, the following inequalities with powers of n, powers of 2, and factorial values are also true:

```
O(n**2) < O(n**3) < O(2**n) < O(n!)
```

If you are unsure about any of the preceding inequalities, perform an online search for tutorials that provide the necessary details.

TASK: MAXIMUM AND MINIMUM POWERS OF AN INTEGER

The code sample in this section shows you how to calculate the largest (smallest) power of a number num whose base is k that is less than (greater than) num, where num and k are both positive integers.

For example, 16 is the largest power of two that is *less* than 24 and 32 is the smallest power of two that is *greater* than 24. As another example, 625 is the largest power of five that is *less* than 1000 and 3125 is the smallest power of five that is *greater* than 1000.

Listing 3.1 displays the contents of `max_min_power_k2.py` that illustrates how to calculate the largest (smallest) power of a number whose base is k that is less than (greater than) a given number. Just to be sure that the task is clear: num and k are positive integers, and the purpose of this task is two-fold:

- find the *largest* number powk such that `k**powk <= num`
- find the *smallest* number powk such that `k**powk >= num`

LISTING 3.1: min_max_power_k2.py

```
def min_max_powerk(num,k):
  powk = 1
  while(powk <= num):
    powk *= k
  if(powk > num):
    powk /= k
  return int(powk), int(powk*k)

nums = [24,17,1000]
powers = [2,3,4,5]

for num in nums:
  for k in powers:
    lowerk,upperk = min_max_powerk(num, k)
    print("num:",num,"lower",lowerk,"upper:",upperk)
  print()
```

Listing 3.1 starts with the function `max_min_powerk()` that contains a loop that repeatedly multiplies the local variable powk (initialized with the value 1) by k. When powk exceeds the parameter num, then powk is divided by k so that we have the lower bound solution.

Note that this function returns powk and powk*k because this pair of numbers is the lower bound and higher bound solutions for this task. Launch the code in Listing 3.1 and you will see the following output:

```
num: 24 lower 16 upper: 32
num: 24 lower 9 upper: 27
num: 24 lower 16 upper: 64
num: 24 lower 5 upper: 25

num: 17 lower 16 upper: 32
num: 17 lower 9 upper: 27
num: 17 lower 16 upper: 64
num: 17 lower 5 upper: 25

num: 1000 lower 512 upper: 1024
num: 1000 lower 729 upper: 2187
num: 1000 lower 256 upper: 1024
num: 1000 lower 625 upper: 3125
```

TASK: BINARY SUBSTRINGS OF A NUMBER

Listing 3.2 displays the contents of the `binary_numbers.py` that illustrates how to display all binary substrings whose length is less than or equal to a given number.

LISTING 3.2: binary_numbers.py

```
import numpy as np

def binary_values(width):
  print("=> binary values for width=",width,":")
  for i in range(0,2**width):
    bin_value = bin(i)
    str_value = str(bin_value)
    print(str_value[2:])
  print()


max_width = 4
for ndx in range(1,max_width):
  binary_values(ndx)
```

Listing 3.2 starts with the function `binary_values()` whose loop iterates from 0 to `2**width`, where `width` is the parameter for this function. The loop variable is `i` and during each iteration, `bin_value` is initialized with the binary value of `i`.

Next, the variable `str_value` is the string-based value of `bin_value`, which is stripped of the two leading characters `0b`. Now launch the code in Listing 3.2 and you will see the following output:

```
=> binary values for width= 1 :
0
1

=> binary values for width= 2 :
0
1
10
11

=> binary values for width= 3 :
0
1
10
11
100
101
110
111
```

TASK: COMMON SUBSTRING OF TWO BINARY NUMBERS

Listing 3.3 displays the contents of common_bits.py that illustrates how to find the longest common substring of two binary strings.

LISTING 3.3: common_bits.py

```
def common_bits(num1, num2):
    bin_num1 = bin(num1)
    bin_num2 = bin(num2)
    bin_num1 = bin_num1[2:]
    bin_num2 = bin_num2[2:]

    if(len(bin_num2) < len(bin_num1)):
      while(len(bin_num2) < len(bin_num1)):
        bin_num2 = "0" + bin_num2

    print(num1,"=",bin_num1)
    print(num2,"=",bin_num2)

    common_bits2 = 0
    for i in range(0,len(bin_num1)):
      if((bin_num1[i] == bin_num2[i]) and (bin_num1[i]
=='1')):
          common_bits2 += 1
    return common_bits2

nums1 = [61,28, 7,100,189]
nums2 = [51,14,28,110, 14]

for idx in range(0,len(nums1)):
  num1 = nums1[idx]
  num2 = nums2[idx]
  common = common_bits(num1, num2)

  print(num1,"and",num2,"have",common,"bits in common")
  print()
```

Listing 3.3 starts with the function common_bits() that initializes the binary numbers bin_num1 and bin_num2 with the binary values of the two input parameters, after which the initial string 0b is removed from both numbers.

Next, a loop iterates from 0 to the length of the string bin_num1 in order to check each digit to see whether or not it equals 1. Each time that the digit 1 is found, the value of common_bits2 (initialized with the value 0) is incremented. When the loop terminates, the variable common_bits2 equals the number of times that bin_num1 and bin_num2 have a 1 in the same position.

The final portion of Listing 3.3 iterates through a pair of arrays with positive integers values and invokes common_bits() during each iteration of the loop. Now launch the code in Listing 3.3 and you will see the following output:

```
61 = 111101
51 = 110011
61 and 51 have 3 bits in common

28 = 11100
14 = 01110
28 and 14 have 2 bits in common

7 = 111
28 = 11100
7 and 28 have 3 bits in common

100 = 1100100
110 = 1101110
100 and 110 have 3 bits in common

189 = 10111101
14 = 00001110
189 and 14 have 2 bits in common
```

TASK: MULTIPLY AND DIVIDE VIA RECURSION

Listing 3.4 displays the contents of the `recursive_multiply.py` that illustrates how to compute the product of two positive integers via recursion.

LISTING 3.4: recursive_multiply.py

```python
import numpy as np

def add_repeat(num, times, sum):
  if(times == 0):
    return sum
  else:
    return add_repeat(num, times-1, num+sum)

arr1 = np.array([5,13,25,17,100])
arr2 = np.array([9,10,25,10,100])

for i in range(0,len(arr1)):
  num1 = arr1[i]
  num2 = arr2[i]
  prod = add_repeat(num1, num2, 0)
  print("product of",num1,"and",num2,"=",prod)
```

Listing 3.4 starts with the function `add_repeat(num,times,sum)` that performs repeated addition by recursively invokes itself. Note that this function uses tail recursion: each invocation of the function replaces times with times-1 and also replaces sum with num+sum (the latter is the tail recursion).

The terminating condition is when times equals 0, at which point the function returns the value of sum. Now launch the code in Listing 3.4 and you will see the following output:

```
product of 5 and 9 = 45
product of 13 and 10 = 130
product of 25 and 25 = 625
product of 17 and 10 = 170
product of 100 and 100 = 10000
```

Listing 3.5 displays the contents of the recursive_divide.py that illustrates how to compute the quotient of two positive integers via recursion.

LISTING 3.5: recursive_divide.py

```python
import numpy as np

def sub_repeat(num1, num2, remainder):
  if(num1 < num2):
    return num1
  else:
    #print("num1-num2:",num1-num2,"num2:",num2)
    return sub_repeat(num1-num2, num2, remainder)

arr1 = np.array([9,13,25,17,100])
arr2 = np.array([5,10,25,10,100])

for i in range(0,len(arr1)):
  num1 = arr1[i]
  num2 = arr2[i]
  prod = sub_repeat(num1, num2, 0)
  print("remainder of",num1,"/",num2,"=",prod)
```

Listing 3.5 contains code that is very similar to Listing 3.3: the difference involves replacing addition with subtraction. Launch the code in Listing 3.5 and you will see the following output:

```
remainder of 9 / 5 = 4
remainder of 13 / 10 = 3
remainder of 25 / 25 = 0
remainder of 17 / 10 = 7
remainder of 100 / 100 = 0
```

TASK: SUM OF PRIME AND COMPOSITE NUMBERS

Listing 3.6 displays the contents of the pair_sum_sorted.py that illustrates how to determine whether or not a sorted array contains the sum of two specified numbers.

LISTING 3.6: pair_sum_sorted.py

```
import numpy as np

PRIME_NUM = 1
COMPOSITE = 0
prime_sum = 0
comp_sum  = 0
prime_list = np.array([])
comp_list  = np.array([])
arr1 = np.array([5,10,17,23,30,47,50])

def is_prime(num):
  div = 2

  while(div < num):
   if( num % div != 0):
      div += 1
   else:
      return COMPOSITE
   return PRIME_NUM

for ndx in range(0,len(arr1)):
  num = arr1[ndx]

  if(is_prime(num) == PRIME_NUM):
    prime_list = np.append(prime_list, num)
    prime_sum += num
  else:
    comp_list = np.append(comp_list, num)
    comp_sum += num

print("prime list:",prime_list)
print("comp  list:",comp_list)
print("prime sum: ",prime_sum)
print("comp sum:  ",comp_sum)
```

Listing 3.6 starts with the function is_prime() that determines whether or not its input parameter is a prime number. The next portion of code in Listing 3.6 is a loop that ranges from 0 to the number of elements. During each iteration, the current number is added to the variable prime_sum if that number is a prime; otherwise, it is added to the variable comp_sum. The final portion of Listing 3.6 displays the sum of the even numbers and the sum of the odd numbers in the input array arr1. Launch the code in Listing 3.6 and you will see the following output:

```
prime list: [ 5. 17. 23. 47.]
comp  list: [10. 30. 50.]
prime sum:   92
comp sum:    90
```

The next portion of this chapter contains various examples of string-related tasks. If need be, you can review the relevant portion of Chapter 1 regarding some of the `Python` built-in string functions, such as `int()` and `len()`.

TASK: COUNT WORD FREQUENCIES

Listing 3.7 displays the contents of the `word_frequency.py` that illustrates how to determine the frequency of each word in an array of sentences.

LISTING 3.7: word_frequency.py

```python
import numpy as np

def word_count(words,check_word):
  count = 0
  for word in words:
    if(word.lower() == check_word.lower()):
      count += 1
  return count

sents = np.array([["I", "love", "thick", "pizza"],
                  ["I", "love", "deep", "dish","pizza"],
                  ["Pepperoni","and","sausage","pizza"],
                  ["Pizza", "with", "mozzarrella"]],dtype=object)

words = np.array([])
for sent in sents:
  for word in sent:
    words = np.append(words,word)

word_counts = {}
for word in words:
  count = word_count(words,word)
  word_counts[word] = count

print("word_counts:")
print(word_counts)
```

Listing 3.7 starts with the function `word_count()` that counts the number of occurrences of a given word in a sentence. The next portion of Listing 3.7 contains a loop that iterates through each sentence of an array of sentences. For each sentence, the code invokes the function `word_count()` with each word in the current sentence. Launch the code in Listing 3.7 and you will see the following output:

```
word_counts:
{'I': 2, 'love': 2, 'thick': 1, 'pizza': 4, 'deep': 1,
'dish': 1, 'Pepperoni': 1, 'and': 1, 'sausage': 1, 'Pizza':
4, 'with': 1, 'mozzarrella': 1}
```

Listing 3.8 displays the contents of the `word_frequency2.py` that illustrates another way to determine the frequency of each word in an array of words.

LISTING 3.8: word_frequency2.py

```
import numpy as np

sents = np.array([["I", "love", "thick", "pizza"],
                  ["I", "love", "deep", "dish","pizza"],
                  ["Pepperoni","and","sausage","pizza"],
                  ["Pizza", "with", "mozzarrella"]],dtype=object)

word_counts = dict()
for sent in sents:
  for word in sent:
    word = word.lower()
    print("word:",word)

    if(word not in word_counts.keys()):
        word_counts[word] = 0
    word_counts[word] += 1

print("word_counts:")
print(word_counts)
```

Listing 3.8 is similar to Listing 3.7, with the following difference: the former contains a simple loop that populates a `Python` dictionary `word_counts` with word frequencies whereas the latter contains a *nested* loop to accomplish the same task. Launch the code in Listing 3.8 and you will see the following output:

```
word_counts:
{'i': 2, 'love': 2, 'thick': 1, 'pizza': 4, 'deep': 1, 'dish': 1,
'pepperoni': 1, 'and': 1, 'sausage': 1, 'with': 1, 'mozzarrella': 1}
```

TASK: CHECK IF A STRING CONTAINS UNIQUE CHARACTERS

The solution involves keeping track of the number of occurrences of each `ASCII` character in a string, and returning `False` if that number is greater than 1 for any character (otherwise return `True`). Therefore, one constraint for this solution is that it's restricted to Indo-European languages that do not have accent marks.

Listing 3.9 displays the contents of the `unique_str.py` that illustrates how to determine whether or not a string contains unique letters.

LISTING 3.9: unique_chars.py

```
import numpy as np

def unique_chars(str):
  if (len(str) > 128):
    return false
```

```
str = str.lower()

char_set = np.zeros([128])

for i in range (0,len(str)):
  char = str[i]
  val = ord('z') - ord(char)
  #print("val:",val)

  if (char_set[val] == 1):
    # found duplicate character
    return False
  else:
    char_set[val] = 1

return True

arr1 = np.array(["a string", "second string", "hello
world"])

for str in arr1:
  print("string:",str)
  result = unique_chars(str)
  print("unique:",result)
  print()
```

Listing 3.9 starts with the function unique_chars() that converts its parameter str to lower case letters and then initializes the 1x128 integer array char_set whose values are all 0. The next portion of this function iterates through the characters of the string str and initializes the integer variable val with the offset position of each character from the character z.

If this position in char_set equals 1, then a duplicate character has been found; otherwise, this position is initialized with the value 1. Note that the value False is returned if the string str contains duplicate letters, whereas the value True is returned if the string str contains unique characters. Now launch the code in Listing 3.9 and you will see the following output:

```
string: a string
unique: True

string: second string
unique: False

string: hello world
unique: False
```

TASK: INSERT CHARACTERS IN A STRING

Listing 3.10 displays the contents of the `insert_chars.py` that illustrates how to insert each character of one string in every position of another string.

LISTING 3.10: insert_chars.py

```
def insert_char(str1, chr):
  result = str1

  result = chr + str1
  for i in range(0,len(str1)):
    left = str1[:i+1]
    right = str1[i+1:]
    #print("left:",left,"right:",right)
    inserted = left + chr + right

    result = result + " " + inserted
  return result

str1 = "abc"
str2 = "def"
print("str1:",str1)
print("str2:",str2)

insertions = ""
for i in range(0,len(str2)):
  new_str = insert_char(str1, str2[i])
  #print("new_str:",new_str)
  insertions = insertions+ " " + new_str

print("result:",insertions)
```

Listing 3.10 starts with the function `insert_char()` that has a string `str1` and a character `chr` as input parameters. The next portion of code is a loop whose loop variable is `i`, which is used to split the string `str1` into two strings: the left substring from positions 0 to `i`, and the right substring from position `i+1`. A new string with three components is constructed: the left string, the character `chr`, and the right string.

The next portion of Listing 3.10 contains a loop that iterates through each character of `str2`; during each iteration, the code invokes `insert_char()` with string `str1` and the current character. Launch the code in Listing 3.10 and you will see the following output:

```
str1: abc
str2: def
result:  dabc adbc abdc abcd eabc aebc abec abce fabc afbc abfc abcf
```

TASK: STRING PERMUTATIONS

There are several ways to determine whether or not two strings are permutations of each other. One way involves sorting the strings alphabetically: if the resulting strings are equal, then they are permutations of each other.

A second technique is to determine whether or not they have the same number of occurrences for each character. A third way is to add the numeric counterpart of each letter in the string: if the numbers are equal and the strings have the same length, then they are permutations of each other.

Listing 3.11 displays the contents of the string_permute.py that illustrates how to determine whether or not two strings are permutations of each other.

LISTING 3.11: string_permute.py

```python
import numpy as np

def permute(str1,str2):
  str1d = sorted(str1)
  str2d = sorted(str2)
  permute = (str1d == str2d)

  print("string1: ",str1)
  print("string2: ",str2)
  print("permuted:",permute)
  print()

strings1 = ["abcdef", "abcdef"]
strings2 = ["efabcf", "defabc"]

for idx in range(0,len(strings1)):
  str1 = strings1[idx]
  str2 = strings2[idx]
  permute(str1,str2)
```

Listing 3.11 starts with the function permute() that takes the two strings str1 and str2 as parameters. Next, the strings str1d and str2d are initialized with the result of sorting the characters in the strings str1 and str2, respectively. At this point, we can determine whether or not str1 and str2 are permutations of each other by determining whether or not the two strings str1d and str2d are equal. Launch the code in Listing 3.11 and you will see the following output:

```
string1:  abcdef
string2:  efabcf
permuted: False

string1:  abcdef
string2:  defabc
permuted: True
```

TASK: FIND ALL SUBSETS OF A SET

Listing 3.12 displays the contents of the powerset.py that illustrates how to list all the subsets of a set.

LISTING 3.12: powerset.py

```
import numpy as np

# strings of the form:
# [a0, a1, ..., an]
def create_array(width):
  arr1 = np.array([])
  for i in range(0,width):
    str1 = "a"+str(i)
    arr1 = np.append(arr1,str1)
  return arr1

def binary_values(arr1,width):
  print("=> binary values for width=",width,":")
  for num in range(0,2**width):
    bin_value = bin(num)
    str_value = bin_value[2:]

    # left-pad with "0" characters:
    for i in range(0,width-len(str_value)):
      str_value = "0" + str_value

    subset = ""
    # check for '1' in a right-to-left loop:
    for ndx in range(len(str_value)-1,-1,-1):
      chr = str_value[ndx]
      if(chr == '1'):
        subset = subset + " " +arr1[ndx]

    if(subset == ""):
      print("{}")
    else:
      if(subset[0] == " "):
        subset = subset[1:]
      print(subset)

width = 4
arr1 = create_array(width)
print("arr1:",arr1)
binary_values(arr1,width)
```

Listing 3.12 starts with the function create_array that creates (and eventually returns) an array arr1 whose values are of the form [a0, a1, ..., an], where n equals the value of the parameter width.

The next portion of Listing 3.12 is the function binary_values with the parameters arr1 and width. This function contains a loop whose loop variable num iterates from 0 to 2**width.

During each iteration of the loop, a binary version bin_value of the variable num is generated. Next, the variable str_value is initialized with the contents of bin_value, starting from index 2 so that the left-most pair of characters 0b are excluded.

The next portion of Listing 3.12 contains loop in which `str_value` right-padded with 0 until its width equals the value `width`. Listing 3.12 initializes the variable `subset` as an empty string, followed by yet another loop during which an element of the powers is created.

The key idea involves iterating through the contents of `str_value`, and each time that the value 1 is found, update the string subset as follows:

```
subset = subset + " " +arr1[ndx]
```

When the loop has completed execution, print the string { } if subset is an empty string; otherwise, print the contents of subset (and skip any initial whitespace).

The final portion of Listing 3.12 invokes the function `binary_values` with an array whose width is the value `width`, along with an array `arr1` of labeled entries of the form [a0, a1, . . ., an]. Now launch the code in Listing 3.12 and you will see the following output:

```
arr1: ['a0' 'a1' 'a2' 'a3']
=> binary values for width= 4 :
{ }
a3
a2
a3 a2
a1
a3 a1
a2 a1
a3 a2 a1
a0
a3 a0
a2 a0
a3 a2 a0
a1 a0
a3 a1 a0
a2 a1 a0
a3 a2 a1 a0
```

TASK: CHECK FOR PALINDROMES

One way to determine whether or not a string is a palindrome is to compare the string with the reverse of the string: if the two strings are equal, then the string is a palindrome. Moreover, there are two ways to reverse a string: one way involves the `Python reverse()` function, and another way is to process the characters in the given string in a right-to-left fashion, and to append each character to a new string.

Another technique involves iterate through the characters in a left-to-right fashion and compare each character with its corresponding character that is based on iterating through the string in a right-to-left fashion.

Listing 3.13 displays the contents of the `palindrome1.py` that illustrates how to determine whether or not a string or a positive integer is a palindrome.

LISTING 3.13: palindrome1.py

```python
import numpy as np

def palindrome1(str):
  full_len = int(len(str))
  half_len = int(len(str)/2)

  for i in range (0,half_len):
    lchar = str[i]
    rchar = str[full_len-1-i]
    if(lchar != rchar):
      return False
  return True

arr1 = np.array(["rotor", "tomato", "radar","maam"])
arr2 = list([123, 12321, 555])

# CHECK FOR STRING PALINDROMES:
for str in arr1:
  print("check string:",str)
  result = palindrome1(str)
  print("palindrome:   ",result)
  print()

# CHECK FOR NUMERIC PALINDROMES:
for num in arr2:
  print("check number:",num)
  str1 = np.str(num)
  str2 = ""
  for digit in str1:
    str2 += digit

  result = palindrome1(str2)
  print("palindrome:   ",result)
  print()
```

Listing 3.13 starts with the function `palindrome1()` with parameter `str` that is a string. This function contains a loop that starts by comparing the left-most character with the right-most character of the string `str`. The next iteration of the loop advances to the second position of the left-side of the string, and compares that character with the character whose position is second from the right end of the string. This step-by-step comparison continues until the middle of the string is reached. During each iteration of the loop, the value `False` is returned if the pair of characters is different. If all pairs of characters are equal, then the string must be a palindrome, in which case the value `True` is returned.

The next portion of Listing 3.13 contains an array `arr1` of strings and an array `arr2` of positive integers. Next, another loop iterates through the elements of `arr1` and invokes the `palindrome1` function to determine whether or not the current element of `arr1` is a palindrome. Similarly, a second loop iterates through the elements of `arr2` and invokes the `palindrome1` function to determine whether or not the current element of `arr2` is a palindrome. Launch the code in Listing 3.13 and you will see the following output:

```
check string: rotor
palindrome:   True

check string: tomato
palindrome:   False

check string: radar
palindrome:   True

check string: maam
palindrome:   True

check number: 123
palindrome:   False

check number: 12321
palindrome:   True

check number: 555
palindrome:   True
```

TASK: CHECK FOR THE LONGEST PALINDROME

This section extends the code in the previous section by examining substrings of a given string. Listing 3.14 displays the contents of the longest_ palindrome.py that illustrates how to determine the longest palindrome in a given string. Note that a single character is always a palindrome, which means that every string has a substring that is a palindrome (in fact, any single character in any string is a palindrome).

LISTING 3.14: longest_palindrome.py

```
import numpy as np

def check_string(str):
  result = 0
  str_len  = len(str)
  str_len2 = int(len(str)/2)

  for i in range(0,str_len2):
    if(str[i] != str[str_len-i-1]):
      result = 1
      break

  if(result == 0):
    #print(str, "is a palindrome")
```

```
      return str
  else:
    #print(str, "is not a palindrome")
    return None

my_strings = ["abc","abb","abccba","azaaza","abcdefgabccb
ax"]
max_pal_str = ""
max_pal_len = 0

for my_str in my_strings:
  max_pal_str = ""
  max_pal_len = 0
  for i in range(0,len(my_str)-1):
    for j in range(0,len(my_str)-i+1):
      sub_str = my_str[i:i+j]
      #print("checking:",sub_str,"in =>",my_str)
      a_str = check_string(sub_str)

      if(a_str != None):
        if(max_pal_len < len(a_str)):
          max_pal_len = len(a_str)
          max_pal_str = a_str

  print("string:",my_str)
  print("maxpal:",max_pal_str)
  print()
```

Listing 3.14 contains logic that is very similar to Listing 3.13. However, the main difference is that there is a loop that checks if *substrings* of a given string are palindromes. The code also keeps track of the longest palindrome and then prints its value and its length when the loop finishes execution.

Note that it's possible for a string to contain multiple palindromes of maximal length: the code in Listing 3.14 finds only the first such palindrome. However, it might be a good exercise to modify the code in Listing 3.14 to find all palindromes of maximal length. Now launch the code in Listing 3.14 and you will see the following output:

```
string: abc
maxpal: a

string: abb
maxpal: bb

string: abccba
maxpal: abccba

string: azaaza
maxpal: azaaza

string: abcdefgabccbax
maxpal: abccba
```

WORKING WITH SEQUENCES OF STRINGS

This section contains Python code samples that search strings to determine the following:

- the maximum length of a sequence of consecutive 1s in a string
- a given sequence of characters in a string
- the maximum length of a sequence of unique characters

After you complete this section, you can explore variations of these tasks that you can solve using the code samples in this section.

The Maximum Length of a Repeated Character in a String

Listing 3.15 displays the contents of max_char_sequence.py that illustrates how to find the maximal length of a repeated character in a string.

LISTING 3.15: max_char_sequence.py

```python
import numpy as np

def max_seq(my_str,char):
  max = 0
  left = 0
  right = 0
  counter = 0
  for i in range(0,len(my_str)):
    curr_char = my_str[i]
    if(curr_char == char):
      counter += 1
      right = i
      if(max < counter):
        max = counter
        #print("new max:",max)
    else:
      counter = 0
      left = i
      right = i

  print("my_str:",my_str)
  print("max sequence of",char,":",max)
  print()

str_list = np.array(["abcdef","aaaxyz","abcdeeefghij"])
char_list = np.array(["a","a","e"])

for idx in range(0,len(str_list)):
  my_str = str_list[idx]
  char = char_list[idx]
  max_seq(my_str,char)
```

Listing 3.15 starts with the function `max_seq()` whose parameters are a string `my_str` and a character `char`. This function contains a loop that iterates through each character of `my_str` and performs a comparison with `char`. As long as the characters equal `char`, the value of the variables `right` and counter are incremented: `right` represents the right-most index and `counter` contains the length of the substring containing the same character.

However, if a character in `my_str` differs from `char`, then counter is reset to 0, and `left` is reset to the value of `right`, and the comparison process begins anew. When the loop has completed execution, the variable `counter` equals the length of the longest substring consisting of equal characters.

The next portion of Listing 3.15 initializes the array `str_list` that contains a list of strings and the array `char_list` with a list of characters. The final loop iterates through the elements of `str_list` and invokes the function `max_seq()` with the current string and the corresponding character in the array `char_list`. Now launch the code in Listing 3.15 and you will see the following output:

```
my_str: abcdef
max sequence of a : 1

my_str: aaaxyz
max sequence of a : 3

my_str: abcdeeefghij
max sequence of e : 3
```

Find a Given Sequence of Characters in a String

Listing 3.16 displays the contents of `max_substr_sequence.py` that illustrates how to find the right-most substring that matches a given string.

LISTING 3.16: max_substr_sequence.py

```python
import numpy as np

def rightmost_substr(my_str,substr):
  left = -1
  len_substr = len(substr)

  # check for substr from right to left:
  for i in range(len(my_str)-len_substr,-1,-1):
    curr_str = my_str[i:i+len_substr]
    #print("checking curr_str:",curr_str)

    if(substr == curr_str):
      left = i
      break

  if(left >= 0):
    print(substr,"is in index",left,"of:",my_str)
  else:
```

```
    print(substr,"does not appear in",my_str)
  print()

str_list = np.array(["abcdef","aaaxyz","abcdeeefghij"])
substr_list = np.array(["bcd","aaa","cde"])

for idx in range(0,len(str_list)):
  my_str = str_list[idx]
  substr = substr_list[idx]
  print("checking:",substr,"in:",my_str)
  rightmost_substr(my_str,substr)
```

Listing 3.16 starts with the function `rightmost_substr` whose parameters are a string `my_str` and a substring `sub_str`. This function contains a loop that performs a right-most comparison of `my_str` and `sub_str`, and iteratively moves leftward one position until the loop reaches the first index position of the string `my_str`.

After the loop has completed its execution, the variable `left` contains the index position at which there is a match between `my_str` and `sub_str`, and its value will be non-negative. If there is no matching substring, then the variable `left` will retain its initial value of -1. In either case, the appropriate message is printed. Now launch the code in Listing 3.16 and you will see the following output:

```
checking: bcd in: abcdef
bcd is in index 1 of: abcdef

checking: aaa in: aaaxyz
aaa is in index 0 of: aaaxyz

checking: cde in: abcdeeefghij
cde is in index 2 of: abcdeeefghij
```

TASK: LONGEST SEQUENCES OF SUBSTRINGS

This section contains `Python` code samples that search strings to determine the following:

- the longest subsequence of unique characters in a given string
- the longest subsequence that is repeated in a given string

After you complete this section you can explore variations of these tasks that you can solve using the code samples in this section.

The Longest Sequence of Unique Characters

Listing 3.17 displays the contents of `longest_unique.py` that illustrates how to find the longest sequence of unique characters in a string.

LISTING 3.17: longest_unique.py

```python
import numpy as np

def rightmost_substr(my_str):
    left = 0
    right = 0
    sub_str = ""
    longest = ""
    my_dict = dict()

    for pos in range(0,len(my_str)):
        char = my_str[pos]
        if(char not in my_dict.keys()):
            my_dict[char] = 1
            unique = my_str[left:pos+1]
            #print("new unique:", unique)

            if(len(longest) < len(unique)):
                longest = unique
                right = pos
        else:
            my_dict = dict()
            left = pos+1
            right = pos+1

    print("longest unique:",longest)
    print()

str_list = np.array(["abcdef","aaaxyz","abcdeeefghij"])

for idx in range(0,len(str_list)):
    my_str = str_list[idx]
    print("checking:",my_str)
    rightmost_substr(my_str)
```

Listing 3.17 starts with the function `rightmost_substr` whose parameter is a string `my_str`. This function contains a right-to-left loop and stores the character in the current index position in the dictionary `my_dict`. If the character has already been encountered, then it's a duplicate character, at which point the length of the current substring is compared with the length of the longest substring that has been found thus far, at which point the variable `longest` is updated with the new value. In addition, the left position `left` and the right position `right` are reset to `pos+1`, and the search for a unique substring begins anew.

After the loop has completed its execution, the value of the variable `longest` equals the length of the longest substring of unique characters.

The next portion of Listing 3.17 initializes the variable `str_list` as an array of strings, followed by a loop that iterates through the elements of `str_list`. The function `rightmost_substr()` is invoked during each iteration in order to find the longest unique substring of the current string. Now launch the code in Listing 3.17 and you will see the following output:

```
checking: abcdef
longest unique: abcdef

checking: aaaxyz
longest unique: axyz

checking: abcdeeefghij
longest unique: efghij
```

The Longest Repeated Substring

Listing 3.18 displays the contents of max_repeated_substr.py that illustrates how to find the longest substring that is repeated in a given string.

LISTING 3.18: max_repeated_substr.py

```
def check_string(my_str, sub_str, pos):
  str_len = len(my_str)
  sub_len = len(sub_str)
  #print("my_str:",my_str,"sub_str:",sub_str)
  match = None
  part_str = ""
  left = 0
  right = 0

  for i in range(0,str_len-sub_len-pos):
    left  = pos+sub_len+i
    right = left+sub_len
    #print("left:",left,"right:",right)
    part_str = my_str[left:right]

    if(part_str == sub_str):
      match = part_str
      break

  return match,left

print("==> Check for repeating substrings of length at
least 2")
my_strings = ["abc","abb","abccba","azaaza","abcdefgabccba
xyz"]

for my_str in my_strings:
  half_len = int(len(my_str)/2)
  max_len = 0
  max_str = ""
  for i in range(0,half_len+1):
    for j in range(2,half_len+1):
      sub_str = my_str[i:i+j]
      a_str,left = check_string(my_str, sub_str,i)
```

```
        if(a_str != None):
          print(a_str,"appears in pos",i,"and pos",left,"in
=>",my_str)
            if(max_len < len(a_str)):
              max_len = len(a_str)
              max_str = a_str

   if(max_str != ""):
     print("=> Maximum repeating substring:",max_str)
   else:
     print("No maximum repeating substring:",my_str)
   print()
```

Listing 3.18 starts with the function check_string() that counts the number of occurrences of the string sub_str in sub_str, starting from index position pos.

This function initializes some scalar values, such as str_len and sub_len that are initialized, respectively, with the length of the search string and the length of a substring. The next portion of this function contains a loop that initializes a string part_str that is a substring of my_str, starting from the value of i (which is the loop variable). If there is a match, the loop terminates and the function returns this matching substring and the left-most position of this substring in the original string.

The second part of Listing 3.18 initializes the variable my_strings with a list of strings to be checked for a repeating substring of maximal length, followed by a triply nested loop that iterates through each string in my_strings. The intermediate loop iterates with the loop variable i from the value 0 to the value half_len+1, which is 1 greater than half the length of the current substring my_str. The innermost loop iterates with the variable j from 2 to half_len+1, in order to initialize the variable sub_str whose contents are the characters from position i through i+j of the string my_str.

At this point the function check_string() is invoked with the string sub_str, and if the returned string is non-empty and has length greater than max_len, then the variables max_len and max_str are updated accordingly. Now launch the code in Listing 3.18 and you will see the following output:

```
==> Check for repeating substrings of length at least 2
No maximum repeating substring: abc

No maximum repeating substring: abb

No maximum repeating substring: abccba

az appears in pos 0 and pos 3 in => azaaza
aza appears in pos 0 and pos 3 in => azaaza
za appears in pos 1 and pos 4 in => azaaza
=> Maximum repeating substring: aza
```

```
ab appears in pos 0 and pos 7 in => abcdefgabccbaxyz
abc appears in pos 0 and pos 7 in => abcdefgabccbaxyz
bc appears in pos 1 and pos 8 in => abcdefgabccbaxyz
=> Maximum repeating substring: abc
```

TASK: MATCH A STRING WITH A WORD LIST (SIMPLE CASE)

This task requires you to tokenize a string into the set of words in a given word list, where multiple matches of the same word are allowed. For example, the string "toeattoeat" can be tokenized by the words in the word list ["to","eat"], where there are two occurrences of "to" as well as "eat" in the string. However, the string "batsit" *cannot* be tokenized by the word list ["bat","it"].

Listing 3.19 displays the contents of multi_word_match.py that illustrates how to solve the simplified version of this task.

LISTING 3.19: multi_word_match.py

```
import numpy as np

# this sequence will not work:
#my_strings = np.array([["bitsandbites"]])
#word_dicts = np.array([["bit","and","bites"]])

# these strings can be tokenized:
my_strings = np.array([["bitsandbites"],["funstuff"],["funst
ufffun"], ["toeattoeattoeat"]])

word_dicts = np.array([["bits","and","bites"],["fun","stuff"],
["fun","stuff"],["to","eat"]])

for idx in range(0, len(my_strings)):
  one_str1 = my_strings[idx]
  the_dict = word_dicts[idx]
  curr_str = one_str1[0]
  matches = np.array([])

  print("=> CURRENT STRING: ",curr_str)
  while((len(curr_str) > 0)):
    for word in the_dict:
      if(curr_str.startswith(word)):
        #print(curr_str, " starts with ",word)
        curr_str = curr_str[len(word):len(curr_str)]
        #print("new curr_str: ", curr_str)

        if(len(matches) == 0):
          matches = np.array([word])
        else:
          matches = np.append(matches,[word])
        #print("NEW match list = ",matches)

      if(len(curr_str) == 0):
```

```
        print("=> FINAL MATCHES: ",matches)
        break
    else:
        print("curr_str = ",curr_str)

  if(len(curr_str) > 0):
    print("Cannot split: ", one_str1[0])
  print("---------------------------\n")
```

Listing 3.19 starts by initializing the variable my_strings as a NumPy array of strings, followed by the variable word_dicts that contains a set of strings that are used to tokenize the strings in the variable my_strings.

The next portion of Listing 3.19 contains a loop that iterates through the elements of the variable my_strings. Several variables are initialized before each iteration, such as the_dict that contains the tokenization tokens that will be used to determine whether or not the current string in my_strings can be fully tokenized.

Next, a while loop executes as long as curr_str has positive length, which is successively reduced by removing a matching token string, starting from index 0, using the startswith() function. Note that the process of matching the tokens is performed by a for loop that is nested inside the while loop. Launch the code in Listing 3.19 and you will see the following output:

```
=> CURRENT STRING:  bitsandbites
curr_str =  andbites
curr_str =  bites
=> FINAL MATCHES:  ['bits' 'and' 'bites']
---------------------------

=> CURRENT STRING:  funstuff
curr_str =  stuff
=> FINAL MATCHES:  ['fun' 'stuff']
---------------------------

=> CURRENT STRING:  funstufffun
curr_str =  stufffun
curr_str =  fun
=> FINAL MATCHES:  ['fun' 'stuff' 'fun']
---------------------------

=> CURRENT STRING:  toeattoeattoeat
curr_str =  eattoeattoeat
curr_str =  toeattoeat
curr_str =  eattoeat
curr_str =  toeat
curr_str =  eat
=> FINAL MATCHES:  ['to' 'eat' 'to' 'eat' 'to' 'eat']
---------------------------
```

The Harder Case

An example of the more difficult case for this task involves a string that cannot be tokenized in a left-to-right fashion, as shown here:

```
#my_strings = np.array([["bitsandbites"]])
#word_dicts = np.array([["bit","and","bites"]])
```

The first pass will remove the string "bit", so the new string to tokenize is now "sandbites". Now you can either use the `find()` function to check which words (if any) in the word list match this new string, and then invoke the `replace()` function to replace the matching word with an empty string.

Notice that there are two words that match the string: "and" as well as "bites". If you match with "and" and remove this word, the new substring is "sbites", which now matches "bites", and the final irreducible string is simple "s". Alternatively, if you match with "bites" and remove this word, the new substring is "sand", which now matches "and", and the final irreducible string is also "s".

Use the information in the preceding paragraph to modify the code in Listing 3.19 (or create another `Python` script) in order to match intermediate locations of a string with words that appear in a word list.

WORKING WITH 1D ARRAYS

A *one-dimensional array* in `Python` is a one-dimensional construct whose elements are homogeneous (i.e., mixed data types are not permitted). Given two arrays A and B, you can add or subtract them, provided that they have the same number of elements. You can also compute the inner product of two vectors by calculating the sum of their component-wise products.

Now that you understand some of the rudimentary operations with one-dimensional matrices, the following subsections illustrate how to perform various tasks on arrays in `Python`.

Rotate an Array

Listing 3.20 displays the contents of the `Python` script `rotate_list.py` that illustrates how to rotate the elements in a list.

LISTING 3.20: rotate_list.py

```
import numpy as np

mylist = [5,10,17,23,30,47,50]
print("original:",mylist)

shift_count = 2
for ndx in range(0,shift_count):
  item = mylist.pop(0)
  arr1 = mylist.append(item)

print("rotated: ",list)
```

Listing 3.20 initializes the variable `mylist` as a list of integers and the variable `shift_count` with the value 2: the latter is the number of positions to shift leftward the elements in `list`. The next portion of Listing 3.20 is a loop that performs two actions: 1) "pop" the left-most element of `mylist` and 2) append that element to `mylist` so that it becomes the new right-most element. The loop is executed `shift_count` iterations, after which the elements in `mylist` have been rotated the specified number of times. Launch the code in Listing 3.20 and you will see the following output:

```
original: [5, 10, 17, 23, 30, 47, 50]
rotated:  [17, 23, 30, 47, 50, 5, 10]
```

TASK: SHIFT NON-ZERO ELEMENTS LEFTWARD

Listing 3.21 displays the contents of `shift_nonzeroes_left.py` that illustrates how to shift non-zero values toward the left while maintaining the relative positions of the non-zero elements.

LISTING 3.21: shift_nonzeroes_left.py

```python
import numpy as np

left=-1
arr1 = np.array([0,10,0,0,30,60,0,200,0])

print("Initial:")
for i in range(0,len(arr1)):
  print(arr1[i],end=" ")
print()

# find the right-most index with value 0:
for i in range(0,len(arr1)):
   if(arr1[i] == 0):
     left = i
   else:
     left += 1
     break

print("non-zero index: ",left)

# ex: 0 10 0 0 30 60 0 200 0
# right shift positions left-through-(idx-1):

for idx in range(left+1,len(arr1)):
  if(arr1[idx] == 0):
    for j in range(idx-1,left,-1):
      arr1[j+1] = arr1[j]

    arr1[left] = 0
    print("shifted non-zero position ",left)
    left += 1
```

```
print("switched:")
for i in range(0,len(arr1)):
  print(arr1[i],end=" ")
print()
```

Listing 3.21 initializes the variable list as a list of integers and the variable shift_count with the value 2: the latter is the number of positions to shift leftward the elements in list. The next portion of Listing 3.21 is a loop that performs two actions:

1. "Pop" the left-most element of list.
2. Append that element to list so that it becomes the new right-most element.

The loop is executed shift_count iterations, after which the elements in list have been rotated the specified number of times. Launch the code in Listing 3.21 and you will see the following output:

```
Initial:
0 10 0 0 30 60 0 200 0
non-zero index:   1
shifted non-zero position   1
shifted non-zero position   2
shifted non-zero position   3
shifted non-zero position   4
switched:
0 0 0 0 0 30 30 60 200
```

TASK: SORT ARRAY IN-PLACE IN O(N) WITHOUT A SORT FUNCTION

Listing 3.22 displays the contents of the Python script simple_sort.py that illustrates a very simple way to sort an array containing an *equal* number of values 0, 1, and 2 without using another data structure.

LISTING 3.22: simple_sort.py

```
arr1 = [0,1,2,2,1,0,0,1,2]
zeroes = 0

print("Initial:")
for i in range(0,len(arr1)):
    print(arr1[i],end=" ")
print()

for i in range(0,len(arr1)):
    if(arr1[i] == 0):
      zeroes += 1

third = int(len(arr1)/3)
for i in range(0,third):
```

```
     arr1[i]         = 0
     arr1[third+i]   = 1
     arr1[2*third+i] = 2

print("Sorted:")
for i in range(0,len(arr1)):
   print(arr1[i],end=" ")
print()
```

Listing 3.22 initializes arr1 with a list of multiple occurrences the values 0, 1, and 2, and then displays the contents of arr1. The second loop counts the number of occurrences the value 0 in the variable arr1. The third loop uses a "thirds" technique to assign the values 0, 1, and 2 to contiguous locations: all the 0 values appear first, followed by all the 1 values, and then all the 2 values. The key word in this task is "equal", which is shown in bold at the top of this section. Launch the code in Listing 3.22 and you will see the following output:

```
Initial:
0 1 2 2 1 0 0 1 2
Sorted:
0 0 0 1 1 1 2 2 2
```

TASK: INVERT ADJACENT ARRAY ELEMENTS

Listing 3.23 displays the contents of the Python script invert_items.py that illustrates how to perform a pairwise inversion of adjacent elements in an array. Specifically, index 0 and index 1 are switched, then index 2 and index 3 are switched, and so forth until the end of the array.

LISTING 3.23: invert_items.py

```
import numpy as np

arr1 = np.array([5,10,17,23,30,47,50])
print("original:",arr1)

mid_point = int(len(arr1)/2)

for temp in range(0,mid_point+2,2):
   temp = arr1[ndx]
   arr1[ndx] = arr1[ndx+1]
   arr1[ndx+1] = temp

print("inverted:",arr1)
```

Listing 3.23 starts with the array arr1 of integers and the variable mid_point that is the mid point of arr1. The next portion of Listing 3.23 contains a loop that iterates from index 0 to index mid_point+2, where the loop variable ndx is incremented by 2 (not by 1) after each iteration. As you can see, the code performs a standard "swap" of the contents of arr1 in index positions ndx and ndx+1 by means of a temporary variable called temp. Launch the code in Listing 3.23 and you will see the following output:

```
original: [ 5 10 17 23 30 47 50]
inverted: [10  5 23 17 47 30 50]
```

Listing 3.24 displays the contents of the Python script swap.py that illustrates how to invert adjacent values in an array without using an intermediate temporary variable. Notice that Listing 3.23 uses a temporary variable temp to switch adjacent values, whereas Listing 3.24 does not require a temporary variable.

LISTING 3.24: swap.py

```
import numpy as np

def swap(num1,num2):
  delta = num2 - num1
  #print("num1:",num1,"num2:",num2)

  num2 = delta
  num1 = num1+delta
  num2 = num1-delta
  #print("num1:",num1,"num2:",num2)
  return num1,num2

arr1 = np.array([15,4,23,35,80,50])
print("BEFORE arr1:",arr1)

for idx in range(0,len(arr1),2):
  num1, num2 = swap(arr1[idx],arr1[idx+1])
  arr1[idx]   = num1
  arr1[idx+1] = num2
  #print("arr1:",arr1)

print("AFTER  arr1:",arr1)
```

Listing 3.24 starts with the function swap that switches the values of two numbers. If this section of code is not clear to you, try manually executing this code with hard-coded values for num1 and num2. The next portion of Listing 3.24 initializes the array arr1 with integers, followed by a loop that iterates through the values of arr1, and invokes the function swap during each iteration. Launch the code in Listing 3.24 and you will see the following output:

```
BEFORE arr1: [15  4 23 35 80 50]
AFTER  arr1: [ 4 15 35 23 50 80]
```

This concludes the portion of the chapter that discusses strings and string-related tasks. The next portion of the chapter contains Python code samples that illustrate how to perform bit-related operations on binary numbers. When you look at the various data structures that are discussed in this book, binary numbers are more closely aligned to strings or one-dimensional vectors than, say, linked lists, queues, or stacks. However, if you do not need to work with binary numbers, you can treat the following section as optional, and you can return to this section at some point in the future if circumstances make it necessary to do so.

TASK: GENERATE 0 THAT IS THREE TIMES MORE LIKELY THAN A 1

This solution to this task is based on the observation that an AND operator with two inputs generates three 0s and a single 1: we only need to randomly generate a 0 and a 1 and supply those random values as input to the AND operator.

Listing 3.25 displays the contents of three_zeroes_and_one.py that illustrates how to generate 0s and 1s with the expected frequency.

LISTING 3.25: three_zeroes_and_one.py

```
import random

# PART 1
COUNT = 10
ones   = 0
zeroes = 0

for idx in range(0,COUNT):
  x = random.randrange(2)
  y = random.randrange(2)
  ans = x & y

  print("input #1: ",x," input #2: ",y, "AND: ",ans)
  if(ans == 1): ones   += 1
  else:             zeroes += 1

print("Percentage of 0s: ",100*zeroes/COUNT)
print("Percentage of 1s: ",100*ones/COUNT)
print("============================\n")

# PART 2
import math
ones   = 0
zeroes = 0
for idx in range(0,COUNT):
  x = int(math.floor(random.random()*2))
  y = int(math.floor(random.random()*2))
  ans = int(x) & int(y)

  print("input #1: ",x," input #2: ",y, "AND: ",ans)
  if(ans == 1): ones   += 1
  else:             zeroes += 1
```

```
print("Percentage of 0s: ",100*zeroes/COUNT)
print("Percentage of 1s: ",100*ones/COUNT)
```

Listing 3.25 starts by initializing several variables, followed by a loop that iterates through the values 0 through COUNT (which is initialized with the value 10). Inside the loop the variables x and y are initialed as random integers that can equal either 0 or 1. Next, the variable ans is initialized as the logical "and" of x and y. If the value of ans is 1 then ones is incremented; otherwise, the variable zeroes is incremented.

In case the logic of the preceding loop is not clear, remember that the logical "and" of two variables that can equal 0 or 1 will be 1 only when the variables are both equal to 1, and in the other three situations the result equals 0. Thus, the the value 0 occurs three times as often as the value 1, as an expected outcome.

Now launch the code in Listing 3.25 and you will see the following output:

```
input #1:  0  input #2:  0 AND:  0
input #1:  1  input #2:  0 AND:  0
input #1:  0  input #2:  1 AND:  0
input #1:  0  input #2:  0 AND:  0
input #1:  0  input #2:  0 AND:  0
input #1:  1  input #2:  1 AND:  1
input #1:  0  input #2:  1 AND:  0
input #1:  0  input #2:  0 AND:  0
input #1:  1  input #2:  1 AND:  1
input #1:  0  input #2:  0 AND:  0
Percentage of 0s:   80.0
Percentage of 1s:   20.0
================================

input #1:  1  input #2:  1 AND:  1
input #1:  1  input #2:  1 AND:  1
input #1:  1  input #2:  1 AND:  1
input #1:  0  input #2:  1 AND:  0
input #1:  1  input #2:  0 AND:  0
input #1:  0  input #2:  0 AND:  0
input #1:  1  input #2:  0 AND:  0
input #1:  1  input #2:  0 AND:  0
input #1:  1  input #2:  0 AND:  0
input #1:  1  input #2:  0 AND:  0
Percentage of 0s:   70.0
Percentage of 1s:   30.0
```

TASK: INVERT BITS IN EVEN AND ODD POSITIONS

This solution to this task involves two parts: the first part "extracts" the bits in the even positions and shifts the result one bit to the right, followed by the second part that extracts the bits in the odd positions and shifts the result one bit to the right. Listing 3.26 displays the contents of swap_adjacent_bits.py that illustrates how to solve this task.

LISTING 3.26: swap_adjacent_bits.py

```python
# swap adjacent bits of a decimal number:
def swap(n):
  return ((n & 0xAAAAAAAA) >> 1) | ((n & 0x55555555) << 1)

arr1 = [70, 1000, 12341234]
for num in arr1:
  print("Decimal: ",num)
  print("Binary:  ",bin(num))
  print("Swapped: ",swap(num))
  print("Swapped: ",bin(swap(num)))
  print("----------------\n")
```

Listing 3.26 defines the Python function swap() that returns the logical "or" of two strings. The first string is the logical "and" of the parameter n and a four-byte string consisting of the hexadecimal value A, after which this result is right-shifted by one bit. The second string is the logical "and" of the parameter n and a four-byte string consisting of the hexadecimal value 5, after which this result is left-shifted by one bit.

The following simplified example might help to understand why the strings in the swap() function are the required values. Consider the decimal value 102 for the parameter n, whose binary value is 1111, and perform the operations in the swap() function, using two-byte hard-coded strings instead of their four-byte counterparts. Since the binary value of 102 equals 01100110, we have the following calculations:

```
n & 0xAA = 0110 0110 & 1010 1010 = 0010 0010
(n & 0XAA) >> 1 = 0010 0010 >> 1 = 0001 0001

n & 0x55 = 0110 0110 & 0101 0101 = 0100 0100
(n & 0x55) << 1 = 0100 & 0100 << 1 = 1000 1000

0001 0001 | 1000 1000 = 1001 1001
```

Compare the preceding code snippet shown in bold with the binary value of 102 and observe that the adjacent bit positions have been switched.

The next portion of Listing 3.26 contains a loop that iterates through the numbers in the list arr1 in order to display their decimal values, as well as their "swapped" values. Now launch the code in Listing 3.26 and you will see the following output:

```
Decimal:  70
Binary:   0b1000110
Swapped:  137
Swapped:  0b10001001
----------------

Decimal:  1000
Binary:   0b1111101000
```

```
Swapped:   980
Swapped:   0b1111010100
-----------------


Decimal:   12341234
Binary:    0b101111000100111111110010
Swapped:   8163313
Swapped:   0b11111001000111111110001
-----------------
```

TASK: INVERT PAIRS OF ADJACENT BITS

This solution is similar to the code in the previous section, which also involves two parts: the "even" part extracts the bits in pairs of adjacent positions, starting from bit positions 2 and 3, and shifts the result two bits to the right, followed by the second part that extracts pairs of bits in the "odd" positions, starting from 0 and 1, and shifts the result two bits to the left. Listing 3.27 displays the contents of swap_adjacent_pairs.py that illustrates how to solve this task.

LISTING 3.27: swap_adjacent_pairs.py

```
def swapAdjacentPairBits(num):
  # even mask: 11001100 => CC
  # odd mask:  00110011 => 33
  return ((num & 0xCCCC) >> 2) | ((num & 0x3333) << 2)

max = 200000
arr1 = [i for i in range(0,max)]

for num in arr1:
  swapped = swapAdjacentPairBits(num)
  bin1 = bin(swapped)[2:]
  dec1 = int(bin1, 2)

  print(f'Binary:  {bin(num):20} decimal: {num:8}')
  print(f'Swapped: {bin(swapped):20} decimal: {dec1:8}')
  print("")
```

Listing 3.27 generalizes the code in Listing 3.26: instead of swapping adjacent bits, the code in this section swaps *adjacent pairs* of bits. Therefore, the Python function swapAdjacentPairBits() specifies the hard-coded strings 0xCCCC and 0x3333 (shown in bold), respectively, instead of 0xAAAA and 0x5555 that are specified in Listing 3.26. Launch the code in Listing 3.27 and you will see the following output:

```
Decimal:  70
Binary:   0b0              decimal:       0
Swapped: 0b0               decimal:       0
```

```
Binary:  0b1                      decimal:        1
Swapped: 0b100                    decimal:        4

Binary:  0b10                     decimal:        2
Swapped: 0b1000                   decimal:        8

Binary:  0b11                     decimal:        3
Swapped: 0b1100                   decimal:       12

Binary:  0b100                    decimal:        4
Swapped: 0b1                      decimal:        1
// output omitted for brevity
Binary:  0b110000110100111011 decimal:     199995
Swapped: 0b11111001110         decimal:       1998

Binary:  0b110000110100111100 decimal:     199996
Swapped: 0b11111000011         decimal:       1987

Binary:  0b110000110100111101 decimal:     199997
Swapped: 0b11111000111         decimal:       1991

Binary:  0b110000110100111110 decimal:     199998
Swapped: 0b11111001011         decimal:       1995

Binary:  0b110000110100111111 decimal:     199999
Swapped: 0b11111001111         decimal:       1999
```

At this point we can generalize this code sample even further to reverse groups of four bits, where the "odd group" is the right-most four bits (with a binary mask of 00001111) and the "even group" is the (left-most four bits (with a binary mask of 11110000): modify the code to switch adjacent triples of bits. Convert the binary masks to their corresponding hexadecimal values, which are 0F and F0, respectively.

Now replace the masks in the swapAdjacentPairsBits() function in Listing 3.27 with the new pair of masks, and also perform left and right shifts of 4 bits instead of 2, as shown here:

```
def swapAdjacentPairBits(num):
  # odd mask:  00001111 => 0F
  # even mask: 11110000 => F0
  return ((num & 0xF0F0) >> 4) | ((num & 0x0F0F) << 4)
```

TASK: FIND COMMON BITS IN TWO BINARY NUMBERS

Listing 3.28 displays the contents of common_bits.py that illustrates how to solve this task.

LISTING 3.28: common_bits.py

```
def common_bits(num1, num2):
    bin_num1 = bin(num1)
    bin_num2 = bin(num2)
```

```
    bin_num1 = bin_num1[2:]
    bin_num2 = bin_num2[2:]

    if(len(bin_num2) < len(bin_num1)):
      while(len(bin_num2) < len(bin_num1)):
        bin_num2 = "0" + bin_num2

    print(num1,"=",bin_num1)
    print(num2,"=",bin_num2)

    common_bits2 = 0
    for i in range(0,len(bin_num1)):
      if((bin_num1[i] == bin_num2[i]) and (bin_num1[i] =='1')):
        common_bits2 += 1
    return common_bits2

nums1 = [61,28, 7,100,189]
nums2 = [51,14,28,110, 14]

for idx in range(0,len(nums1)):
  num1 = nums1[idx]
  num2 = nums2[idx]
  common = common_bits(num1, num2)

  print(num1,"and",num2,"have",common,"bits in common")
  print()
```

Listing 3.28 defines the function common_bits() that initializes the variables bin_num1 and bin_num2 as the binary values for the parameters num1 and num2, respectively. Notice that the first two index positions are skipped: this is necessary in order to exclude the string "0b" that appears in the binary values bin_num1 and bin_num2.

The next portion of Listing 3.28 initializes the lists nums1 and nums2 with lists of positive integers, followed by a loop that iterates through these lists in order to determine the number of bits that are in common to each pair of array values. Now launch the code in Listing 3.28 and you will see the following output:

```
61 = 111101
51 = 110011
61 and 51 have 3 bits in common

28 = 11100
14 = 01110
28 and 14 have 2 bits in common

7 = 111
28 = 11100
7 and 28 have 3 bits in common

100 = 1100100
110 = 1101110
100 and 110 have 3 bits in common
```

```
189 = 10111101
14 = 00001110
189 and 14 have 2 bits in common
```

TASK: CHECK FOR ADJACENT SET BITS IN A BINARY NUMBER

Listing 3.29 displays the contents of check_adjacent_bits.py that illustrates how to solve this task.

LISTING 3.29: check_adjacent_bits.py

```
# true if adjacent bits are set in num:
def check(num):
  return num & (num << 1)

arr1 = [15, 16, 17, 50, 67, 99]
for num in arr1:
  print("Decimal: ",num)
  print("Binary:  ",bin(num))

  if check(num):
    print("Adjacent pair of set bits found")
  else:
    print("No adjacent pair of set bits found")
  print("--------------\n")
```

Listing 3.29 defines the Python function check() that returns the result of the logical "and" of the parameter n with the value (n<<1), which does detect the presence of adjacent bits that are set equal to 1. Now launch the code in Listing 3.29 and you will see the following output:

```
15 in binary =  0b1111
Adjacent pair of set bits found
----------------

16 in binary =  0b10000
No adjacent pair of set bits found
----------------

17 in binary =  0b10001
No adjacent pair of set bits found
----------------

50 in binary =  0b110010
Adjacent pair of set bits found
----------------

67 in binary =  0b1000011
Adjacent pair of set bits found
----------------
99 in binary =  0b1100011
Adjacent pair of set bits found
----------------
```

TASK: COUNT BITS IN A RANGE OF NUMBERS

Listing 3.30 displays the contents of `count_bits.py` that illustrates how to solve this task.

LISTING 3.30: count_bits.py

```
# Given an integer num, return an array of the number of 1's in
# the binary representation of every number in the range [0, num]

def count_bits(num):
  num_bin = bin(num)
  count = 0
  for i in range(0,len(num_bin)):
    if num_bin[i] == '1':
      count += 1
  #print(num,"has",count,"bits")
  return count

number = 20
total_bits = 0
bit_list = list()
for i in range(0,number):
  total_bits += count_bits(i)
  bit_list.append(count_bits(i))

print("=> Array of bit counts for numbers between 0
and",number,":")
print(bit_list)
```

Listing 3.30 defines the function `count_bits()` that initializes the variable `num_bin` as the binary counterpart to the parameter `num`, followed by a loop that counts the number of occurrences of the string '1' in `num_bin`, whose value is returned by this function.

The next portion of Listing 3.30 initializes several variables, including the list variable `bit_list`, followed by a loop that counts the number of bits that appear in the numbers 0 through `number` (which is initialized with the value 20). Now launch the code in Listing 3.30 and you will see the following output:

```
=> Array of bit counts for numbers between 0 and 20 :
[0, 1, 1, 2, 1, 2, 2, 3, 1, 2, 2, 3, 2, 3, 3, 4, 1, 2, 2, 3]
```

TASK: FIND THE RIGHT-MOST SET BIT IN A NUMBER

This solution to this task involves Listing 3.31 displays the contents of the `Python` script `right_most_set_bit.py` that illustrates how to solve this task.

LISTING 3.31: right_most_set_bit.py

```python
import numpy as np
import math

def getFirstSetBitPos(num):
  return math.log2(num & -num)+1

arr1 = np.array([12,18,29,66])
for num in arr1:
  bnum  = bin(num)
  bnum2 = bnum[2:]
  rbit  = (int(getFirstSetBitPos(num)))
  #print(f'num: ",num," bnum: ",bnum," bnum2: ",bnum2," rbit: ",rbit)
  print(f'num: {num:6} bnum: {bnum:10} bnum2: {bnum2:8} rbit:
{rbit:6}')
```

Listing 3.31 defines the function getFirstSetBitPos() that returns 1 plus the logarithm (base 2) of the parameter num and the "and" of (-num). The next portion of Listing 3.31 initializes the variable arr1 as a Numpy array of positive integers, followed by a loop that iterates through each value in arr1. The variable bnum2 is initialized with the binary representation of num (which is an element of arr1), starting from index 2 in order to skip the string "0b".

Next, the variable bit is assigned as the integer result of invoking the function getFirstSetBitPos() with the variable num, and then displays the values of bnum, bnum2, and rbit. Now launch the code in Listing 3.31 and you will see the following output:

```
num:    12 bnum: 0b1100      bnum2: 1100      rbit:     3
num:    18 bnum: 0b10010     bnum2: 10010     rbit:     2
num:    29 bnum: 0b11101     bnum2: 11101     rbit:     1
num:    66 bnum: 0b1000010   bnum2: 1000010   rbit:     2
```

TASK: THE NUMBER OF OPERATIONS TO MAKE ALL CHARACTERS EQUAL

This solution to this task involves Listing 3.32 displays the contents of the Python script flip_bit_count.py that illustrates how to solve this task.

LISTING 3.32: flip_bit_count.py

```python
# determine the minimum number of operations
# to make all characters of the string equal
def minOperations(the_string):
   count = 0; # track the # of changes

   for i in range(1, len(the_string)):
     # are adjacent characters equal?
     if (the_string[i] != the_string[i - 1]):
       count += 1;
```

```
    return(count);

arr1 = [ "0101010101", "1111010101", "100001", "111111"]
for str1 in arr1:
  print("String: ",str1)
  print("Result: ", minOperations(str1));
  print("----------------\n")
```

Listing 3.32 defines the function `minOperations()` that contains a loop to count the number of times that adjacent characters (of a string) are different: this number equals the number of positions that must be "flipped" in order to make all characters equal.

The next portion of Listing 3.32 initials the variable `arr1` as a list of strings that consist of either 0 or 1, followed by a loop that iterates through the elements of `arr1` and then invokes the function `minOperations()` to calculate the number of flips that are required. Now launch the code in Listing 3.32 and you will see the following output:

```
String:   0101010101
Result:   9
----------------


String:   1111010101
Result:   6
----------------


String:   100001
Result:   2
----------------


String:   111111
Result:   0
----------------
```

TASK: COMPUTE XOR WITHOUT XOR FOR TWO BINARY NUMBERS

This solution to this task involves Listing 3.33 displays the contents of the Python script `xor_without_xor.py` that illustrates how to solve this task.

LISTING 3.33: xor_without_xor.py

```
# perform the XOR of two numbers without XOR:
def findBits(x, y):
  return (x | y) - (x & y)


arrx = [65,15]
arry = [80,240]

for idx in range(0,len(arrx)):
  x = arrx[idx]
  y = arry[idx]
```

```
xory  = bin(x|y)
xandy = bin(x&y)

print("Decimal x: ",x)
print("Decimal y: ",y)

print("Binary x:  ",bin(x))
print("Binary y:  ",bin(y))

print("x OR y:    ",xory)
print("x AND y:   ",xandy)
print("x XOR y:   ",bin(findBits(x,y)))
print("------------------\n")
```

Listing 3.33 defines the function `findBits()` that calculates the XOR value of its parameters x and y by computing (and returning) the quantity `(x | y) - (x & y)`, which is logically equivalent to computing the XOR value of x and y. Launch the code in Listing 3.33 and you will see the following output:

```
Decimal x:  65
Decimal y:  80
Binary x:   0b1000001
Binary y:   0b1010000
x OR y:     0b1010001
x AND y:    0b1000000
x XOR y:    0b10001
------------------

Decimal x:  15
Decimal y:  240
Binary x:   0b1111
Binary y:   0b11110000
x OR y:     0b11111111
x AND y:    0b0
x XOR y:    0b11111111
------------------
```

WORKING WITH 2D ARRAYS

A *two-dimensional array* in Python is a two-dimensional construct whose elements are homogeneous (i.e., mixed data types are not permitted). Given two arrays A and B, you can add or subtract them, provided that they have the same number of rows and columns.

Multiplication works differently: if A is an mxn matrix that you want to multiply (on the right of A) by B, then B must be an nxp matrix. The rule for matrix multiplication is as follows: the number of columns of A must equal the number of rows of B.

In addition, the *transpose* of matrix A is another matrix At such that the rows and columns are interchanged. Thus, if A is an mxn matrix then At is an nxm matrix. The matrix A is symmetric if A = At. The matrix A is the identity matrix I if the values in the main diagonal (upper left to lower right) are 1 and the other values are 0. The matrix A is invertible if there is a matrix B such that A*B = B*A = I. Based on the earlier discussion regarding the product of two matrices, both A and B must be square matrices with the same number of rows and columns.

Now that you understand some of the rudimentary operations with matrices, the following subsections illustrate how to perform various tasks on matrices in Python.

THE TRANSPOSE OF A MATRIX

As a reminder, the transpose of matrix A is matrix At, where the rows and columns of A are the columns and rows, respectively, of matrix At.

Listing 3.34 displays the contents of the mat_transpose.py that illustrates how to find the transpose of an mxn matrix.

LISTING 3.34: mat_transpose.py

```
import numpy as np

# the transpose of a matrix is a 90 degree rotation
def transpose(A,rows,cols):
    for i in range(0,rows):
        for j in range(i,cols):
            #print("switching",A[i,j],"and",A[j,i])
            temp = A[i,j]
            A[i,j] = A[j,i]
            A[j,i] = temp
    return A

A = np.array([[100,3],[500,7]])
print("=> original:")
print(A)
At = transpose(A, 2, 2)
print("=> transpose:")
print(At)
print()

# example 2:
A = np.array([[100,3,-1],[30,500,7],[123,456,789]])
print("=> original:")
print(A)
At = transpose(A, 3, 3)
print("=> transpose:")
print(At)
```

Listing 3.34 is actually straightforward: the function transpose() contains a nested loop that uses a temporary variable temp to perform a simple swap of the values of A[i,j] and A[j,i] in order to generate the transpose of the matrix A. The next portion of Listing 3.34 initializes a 2x2 array A and then invokes the function `transpose` to generate its transpose. Launch the code in Listing 3.35 and you will see the following output:

```
=> original:
[[100    3]
 [500    7]]
=> transpose:
[[100 500]
 [  3    7]]

=> original:
[[100    3   -1]
 [ 30 500    7]
 [123 456 789]]
=> transpose:
[[100   30 123]
 [  3 500 456]
 [ -1    7 789]]
```

In case you didn't notice, the transpose `At` of a matrix `A` is actually a 90 degree rotation of matrix `A`. Therefore, if `A` is a square matrix of pixels values for a `PNG`, then `At` is a 90 degree rotation of the `PNG`. However, if you take the transpose of `At`, the result is the original matrix `A`.

SUMMARY

This chapter started with an introduction to one-dimensional vectors, and how to calculate their length and the inner product of pairs of vectors. Then you were taught how to perform various tasks involving numbers, such as multiplying and dividing two positive integers via recursive addition and subtraction, respectively.

In addition, you learned about working with strings; and how to check a string for unique characters, how to insert characters in a string, and how to find permutations of a string. Next, you learned about determining whether or not a string is a palindrome.

Then you learned how to perform bit-related operations to solve various tasks in Python, such as reversing adjacent bits in a binary number, counting the number common occurrences of a 1 in two binary numbers, and how to find the right-most bit that equals 1 in a binary number. Finally, you learned how to calculate the transpose of a matrix, which is the equivalent of rotating a bitmap of an image by 90 degrees.

## CHAPTER 4

SEARCH AND SORT ALGORITHMS

The first half of this chapter provides an introduction to some well-known search algorithms, followed by the second half that discusses various sorting algorithms.

The first section of this chapter introduces search algorithms such as linear search and binary search, that you can use when searching for an item (which can be numeric or character-based) in an array. A linear search is inefficient because it requires an average of n/2 (which has complexity O(n)) comparisons to determine whether or not the search element is in the array, where n is the number of elements in the list or array.

By contrast, a binary search required O (log n) comparisons, which is vastly more efficient with larger sets of items. For example, if an array contains 1,024 items, a most ten comparisons are required in order to find an element because 2**10 = 1024, so log(1024) = 10. However, a binary search algorithm requires a sorted list of items.

The second part of this chapter discusses some well-known sorting algorithms, such as the bubble sort, selection sort, insertion sort, the merge sort, and the quick sort that you can perform on an array of items.

SEARCH ALGORITHMS

The following list contains some well-known search algorithms that will be discussed in several subsections:

- linear search
- binary search
- jump search
- Fibonacci search

A *linear search* algorithm is probably the simplest of all the search algorithms: this algorithm checks every element in an array until either the desired item is located or the end of the array is reached.

However, as you learned in the introduction for this chapter, a linear search is inefficient when an array contains a large number of values. If the array is very small, the difference in performance between a linear search and a binary search can also be very small; in this case, a linear search might be an acceptable choice of algorithms.

In the RDBMS (relational database management system) world, tables often have an index (and sometimes more than one) in order to perform a table search more efficiently. However, there is some additional computational overhead involving the index, which is a separate data structure that is stored on disk. Additionally, a linear search involves only the data in the table. As a rule of thumb, an index-based search is more efficient when tables have more than 300 rows (but results can vary). The next section contains a code sample that performs a linear search on an array of numbers.

Linear Search

Listing 4.1 displays the contents of the `linear_search.py` that illustrates how to perform a linear search on an array of numbers.

LISTING 4.1: linear_search.py

```python
import numpy as np

found = -1
item = 123
arr1 = np.array([1,3,5,123,400])

for i in range(0,len(arr1)):
   if (item == arr1[i]):
      found = i
      break

if (found >= 0):
   print("found",item,"in position",found)
else:
   print(item,"not found")
```

Listing 4.1 starts with the variable `found` that is initialized with the value −1, followed by the search item 123, and also the array `arr1` that contains an array of numbers. Next, a loop that iterates through the elements of the array `arr1` of integers, comparing each element with the value of item. If a match occurs, the variable `found` is set equal to the value of the loop variable `i`, followed by an early exit.

The last portion of Listing 4.1 checks the value of the variable `found`: if it's nonnegative then the search item was found (otherwise it's not in the array). Launch the code in Listing 4.1 and you will see the following output:

```
found 123 in position 3
```

Keep in mind the following point: although the array `arr1` contains a sorted list of numbers, the code works correctly for an unordered list as well.

Binary Search Walk-Through

A *binary search* requires a sorted array and can be implemented via an iterative algorithm as well as a recursive solution. The key idea involves comparing the middle element of an array of sorted elements with a search element. If they are equal, then the item has been found; if the middle element is smaller than the search element then repeat the previous step with the right half of the array; if the middle element is larger than the search element then repeat the previous step with the left half of the array. Eventually the element will be found (if it appears in the array) or the repeated splitting of the array terminates when the subarray has a single element (i.e., no further splitting can be performed).

Let's perform a walk-through of a binary search that searches for an item in a sorted array of integers.

Example #1: let `item` = 25 and `arr1` = [10,20,25,40,100], so the midpoint of the array is 3. Since `arr1[3]` = `item`, the algorithm terminates successfully.

Example #2: let `item` = 25 and `arr1` = [1,5,10, 15, 20, 25, 40], which means that the midpoint is 4.

First iteration: since arr1[4] < item, we search the array [20,25,40]

Second iteration: the midpoint is 1, and the corresponding value is 25.

Third iteration: 25 and the array is the single element [25], which matches the item.

Example #3: let item = 25 and `arr1` = [10, 20, 25, 40, 100,150,400], so the midpoint is 4.

First iteration: since arr1[4] > 25, we search the array [10,20,25].

Second iteration: the midpoint is 1, and the corresponding value is 20.

Third iteration: 25 and the array is the single element [25], which matches the item.

Example #4: item = 25 and `arr1` = [1,5,10, 15, 20, 30, 40], so the midpoint is 4.

First iteration: since arr1[4] < 25, we search the array [20,30,40].

Second iteration: the midpoint is 1, and the corresponding value is 30.

Third iteration: 25 and the array is the single element [20], so there is no match.

As mentioned in the first paragraph of this section, a binary search can be implemented with an interactive solution, which is the topic of the next section.

Binary Search (Iterative Solution)

Listing 4.2 displays the contents of the binary_search.py that illustrates how to perform a binary search with an array of numbers.

LISTING 4.2: binary_search.py

```
import numpy as np

arr1 = np.array([1,3,5,123,400])
left = 0
right = len(arr1)-1
found = -1
item = 123

while(left <= right):
  mid = int(left + (right-left)/2)

  if(arr1[mid] == item):
    found = mid
    break
  elif (arr1[mid] < item):
    left = mid+1
  else:
    right = mid-1

print("array:",arr1)

if( found >= 0):
  print("found",item,"in position",found)
else:
 print(item,"not found")
```

Listing 4.2 initializes an array of numbers and some scalar variables to keep track of the left and right index positions of the subarray that we will search each time that we split the array. The next portion of Listing 4.2 contains conditional logic that implements the sequence of steps that you saw in the examples in the previous section. Launch the code in Listing 4.2 and you will see the following output:

```
array: [  1    3    5 123 400]
found 123 in position 3
```

Binary Search (Recursive Solution)

Listing 4.3 displays the contents of the binary_search_recursive.py that illustrates how to perform a binary search recursively with an array of numbers.

LISTING 4.3: binary_search_recursive.py

```
import numpy as np
def binary_search(data, item, left, right):
```

```
    if left > right:
      return False
    else:
      # incorrect (can result in overflow):
      # mid = (left + right) / 2
      mid = int(left + (right-left)/2)

      if item == data[mid]:
        return True
      elif item < data[mid]:
        # recursively search the left half
        return binary_search(data, item, left, mid-1)
      else:
        # recursively search the right half
        return binary_search(data, item, mid+1, right)
arr1  = np.array([1,3,5,123,400])
items = [-100, 123, 200, 400]

print("array: ",arr1)
for item in items:
  left  = 0
  right = len(arr1)-1
  result = binary_search(arr1, item, left, right)
  print("item:  ",item, " found: ",result)
```

Listing 4.3 start with the function `binary_search()` with parameters `data`, `item`, `left`, and `right`) that contain the current array, the search item, the left index of `data`, and the right index of `data`, respectively. If the left index `left` is greater than the right index `right` then the search item does not exist in the original array.

Contrastingly, if the left index `left` is *less than* the right index `right` then the code assigns the middle index of `data` to the variable `mid`. Next, the code performs the following three-part conditional test:

If `item == data[mid]` then the search item has been found in the array.

If `item < data[mid]` then the function `binary_search()` is invoked with the left-half of the `data` array.

If `item > data[mid]` then the function `binary_search()` is invoked with the right-half of the `data` array.

The next portion of Listing 4.3 initializes the sorted array `arr1` with numbers and initializes the array `items` with a list of search items, and also initializes some scalar variables to keep track of the left and right index positions of the subarray that we will search each time that we split the array.

The final portion of Listing 4.3 consists of a loop that iterates through each element of the items array and invokes the function `binary_search()` to determine whether or not the current item is in the sorted array. Launch the code in Listing 4.3 and you will see the following output:

```
array:   [  1   3   5 123 400]
item:    -100  found:  False
item:    123  found:  True
item:    200  found:  False
item:    400  found:  True
```

WELL-KNOWN SORTING ALGORITHMS

Sorting algorithms have a best case, average case, and worst case in terms of performance. Interestingly, sometimes an efficient algorithm (such as quick sort) can perform the worst when a given array is already sorted.

The following subsections contain code samples for the following well-known sort algorithms:

- bubble sort
- selection sort
- insertion sort
- Merge sort
- Quick sort
- BucketSort
- Shell Sort
- Shell Sort
- Heap Sort
- BucketSort
- InplaceSort
- CountingSort
- RadixSort

If you want to explore sorting algorithms in more depth, perform an Internet search for additional sorting algorithms.

Bubble Sort

A *bubble sort* involves a nested loop whereby each element of an array is compared with the elements to the right of the given element. If an array element is less than the current element, the values are interchanged ("swapped"), which means that the contents of the array will eventually be sorted from smallest to largest value.

Here is an example:

```
arr1 = np.array([40, 10, 30, 20]);
Item = 40;
Step 1: 40 > 10 so switch these elements:
arr1 = np.array([10, 40, 30, 20]);
Item = 40;
Step 2: 40 > 30 so switch these elements:
arr1 = np.array([10, 30, 40, 20]);
Item = 40;
Step 3: 40 > 20 so switch these elements:
arr1 = np.array([10, 30, 20, 40]);
```

As you can see, the smallest element is in the left-most position of the array arr1. Now repeat this process by comparing the second position (which is index 1) with the right-side elements.

```
arr1 = np.array([10, 30, 20, 40]);
Item = 30;
Step 4: 30 > 20 so switch these elements:
arr1 = np.array([10, 20, 30, 40]);
Item = 30;
Step 4: 30 < 40 so do nothing
```

As you can see, the smallest elements two elements occupy the first two positions in the array arr1. Now repeat this process by comparing the third position (which is index 2) with the right-side elements.

```
arr1 = np.array([10, 20, 30, 40]);
Item = 30;
Step 4: 30 < 40 so do nothing
```

The array arr1 is now sorted in increasing order (in a left-to-right fashion). If you want to reverse the order so that the array is sorted in decreasing order (in a left-to-right fashion), simply replace the ">" operator with the "<" operator in the preceding steps.

Listing 4.4 displays the contents of the bubble_sort.py that illustrates how to perform a bubble sort on an array of numbers.

LISTING 4.4: bubble_sort.py

```
import numpy as np
arr1 = np.array([40, 10, 30, 20]);

for i in range(1,arr1.length-1):
  for j in range(i+1,arr1.length):
    if(arr1[i] > arr1[j]):
      temp = arr1[i];
      arr1[i] = arr1[j];
      arr1[j] = temp;
```

You can manually perform the code execution in Listing 4.4 to convince yourself that the code is correct (hint: it's the same sequence of steps that you saw earlier in this section). Launch the code in Listing 4.4 and you will see the following output:

```
initial: [40 10 30 20]
sorted:  [10 20 30 40]
```

Find Anagrams in a List of Words

Recall that the variable word1 is an anagram of word2 if word2 is a permutation of word1. Listing 4.5 displays the contents of the anagrams2.py that illustrates how to check if two words are anagrams of each other.

LISTING 4.5: anagrams2.py

```
def is_anagram(str1, str2):
  sorted1 = sorted(str1)
```

```
    sorted2 = sorted(str2)
    return (sorted1 == sorted2)

words = ["abc","evil","Z","cab","live","xyz","zyx","bac"]
print("=> Initial words:")
print(words)
print()

for i in range(0,len(words)-1):
  for j in range(i+1,len(words)):
    result = is_anagram(words[i], words[j])
    if(result == True):
      print(words[i]," and ",words[j]," are anagrams")
```

Listing 4.5 defines the function is_anagram() that takes parameters str1 and str2 whose sorted values are used to initialize the variables sorted1 and sorted2, respectively. The function returns the result of comparing sorted1 with sorted2: if they are equal then str1 is a palindrome.

The next portion of Listing 4.5 initializes the variable words as a list of strings, followed by a nested loop. The outer loop uses the variable i to range from 0 to len(words)-1, and the inner loop uses the variable j to range from i+1 to len(words). The inner loop initializes the variable result with the value returned by the function is_anagram() that is invoked with the strings words[i] and words[j]. The two words are palindromes if the value of the variable result is True. Launch the code in Listing 4.5 and you will see the following output:

```
=> Initial words:
['abc', 'evil', 'Z', 'cab', 'live', 'xyz', 'zyx', 'bac']

abc   and   cab   are anagrams
abc   and   bac   are anagrams
evil  and   live  are anagrams
cab   and   bac   are anagrams
xyz   and   zyx   are anagrams
```

SELECTION SORT

Listing 4.6 displays the contents of the selection_sort.py that illustrates how to perform a selection sort on an array of numbers.

LISTING 4.6: selection_sort.py

```
import sys

arr1 = [64, 25, 12, 22, 11]

# Traverse through all array elements
for i in range(len(arr1)):
    # Find the minimum element in remaining unsorted array
```

```
  min_idx = i
  for j in range(i+1, len(arr1)):
    if arr1[min_idx] > arr1[j]:
      min_idx = j

  # Swap the found minimum element with the first element
  arr1[i], arr1[min_idx] = arr1[min_idx], arr1[i]

print ("Initial:")
print (arr1)
print ("Sorted: ")
for i in range(len(arr1)):
  print("%d" %arr1[i],end=" ")
print()
```

Listing 4.6 starts by initializing the array `arr1` with some integers, followed by a loop that iterates through the elements of `arr1`. During each iteration of this loop, another inner loop compares the current array element with each element that appears to the right of the current array element. If any of those elements is smaller than the current array element, then the index position of the former is maintained in the variable `min_idx`. After the inner loop has completed execution, the current element is "swapped" with the small element (if any) that has been found via the following code snippet:

```
arr1[i], arr1[min_idx] = arr1[min_idx], arr1[i]
```

In the preceding snippet, `arr1[i]` is the "current" element, and `arr1[min_idx]` is element (to the right of index i) that is smaller than `arr1[i]`. If these two values are the same, then the code snippet swaps `arr1[i]` with itself. Now launch the code in Listing 4.6 and you will see the following output:

```
Initial:
[64, 25, 12, 22, 11]
Sorted:
11 12 22 25 64
```

INSERTION SORT

Listing 4.7 displays the contents of the `insertion_sort.py` that illustrates how to perform a selection sort on an array of numbers.

LISTING 4.7: insertion_sort.py

```
def insertionSort(arr1):
  # Traverse through 1 to len(arr1)
  for i in range(1, len(arr1)):
    key = arr1[i]

    # Move elements of arr1[0..i-1], that are
    # greater than key, to one position ahead
    # of their current position
```

```
        j = i-1
        while j >=0 and key < arr1[j]:
            arr1[j+1] = arr1[j]
            j -= 1
        arr1[j+1] = key
        print("New order:",arr1)

arr1 = [12, 11, 13, 5, 6]
print("Initial:   ", arr1)

insertionSort(arr1)
print ("Sorted:",end=" ")
for i in range(len(arr1)):
  print ("%d" %arr1[i],end=" ")
print()
```

Listing 4.7 starts with the function insertionSort() and contains a loop that iterates through the elements of the array arr1. During each iteration of this loop, the variable key is assigned the value of the element of array arr1 whose index value is the loop variable i. Next, a while loop shift a set of elements to the right of index j, as shown here:

```
j = i-1
while j >=0 and key < arr1[j] :
    arr1[j+1] = arr1[j]
    j -= 1
arr1[j+1] = key
```

For example, after the first iteration of the inner while loop we have the following output:

```
Initial:    [12, 11, 13, 5, 6]
New order: [11, 12, 13, 5, 6]
```

The second iteration of the inner loop does not produce any changes, but the third iteration shifts some of the array elements, at which point we have the following output:

```
Initial:    [12, 11, 13, 5, 6]
New order: [11, 12, 13, 5, 6]
New order: [11, 12, 13, 5, 6]
New order: [5, 11, 12, 13, 6]
```

The final iteration of the outer loop results in an array with sorted elements. Now launch the code in Listing 4.7 and you will see the following output:

```
Initial:    [12, 11, 13, 5, 6]
New order: [11, 12, 13, 5, 6]
New order: [11, 12, 13, 5, 6]
New order: [5, 11, 12, 13, 6]
New order: [5, 6, 11, 12, 13]
Sorted:     5 6 11 12 13
```

COMPARISON OF SORT ALGORITHMS

A bubble sort is rarely used: it's most effective when the data value is already almost sorted. A selection sort is used infrequently: while this algorithm is effective for very short lists, the insertion sort is often superior. An insertion sort is useful if the data are already almost sorted, or if the list is very short (e.g., at most 50 items).

Among the three preceding algorithms, only insertion sort is used in practice, typically as an auxiliary algorithm in conjunction with other more sophisticated sorting algorithms (e.g., quicksort or merge sort).

MERGE SORT

A *merge sort* involves merging two arrays of sorted values. In the following subsections you will see three different ways to perform a merge sort. The first code sample involves a third array, whereas the second and third code samples do not require a third array. Moreover, the third code sample involves one `while` loop whereas the second code sample involves a pair of nested loops, which means that the third code sample is simpler and also more memory efficient.

Merge Sort With a Third Array

The simplest way to merge two arrays involves copying elements from those two arrays to a third array, as shown here:

```
        A              B              C
    +-----+        +-----+        +-----+
    |  20 |        |  50 |        |  20 |    A
    |  80 |        |  70 |        |  50 |    B
    | 200 |   +    | 100 |   =    |  70 |    B
    | 300 |        +-----+        |  80 |    A
    | 500 |                       | 100 |    B
    +-----+                       | 200 |    A
                                  | 300 |    A
                                  | 500 |    A
                                  +-----+
```

The right-most column in the preceding diagram lists the array (either A or B) that contains each number. As you can see, the order ABBABAAA switches between array A and array B. However, the final three elements are from array A because all the elements of array B have been processed.

Two other possibilities exist: array A is processed and B still has some elements, or both A and B have the same size. Of course, even if A and B have the same size, it's still possible that the final sequence of elements is from a single array.

For example, array B is longer than array A in the example below, which means that the final values in array C are from B:

```
A = [20,80,200,300,500]
B = [50,70,100]
```

The following example involves array A and array B with the same length:

```
A = [20,80,200]
B = [50,70,300]
```

The next example also involves array A and array B with the same length, but all the elements of A are copied to B and then all the elements of B are copied to C:

```
A = [20,30,40]
B = [50,70,300]
```

Listing 4.8 displays the contents of the merge_sort1.py that illustrates how to perform a merge sort on two arrays of numbers.

LISTING 4.8: merge_sort1.py

```python
import numpy as np

items1 = np.array([20, 30, 50, 300])
items2 = np.array([80, 100, 200])

def merge_items():
  items3 = np.array([])
  ndx1 = 0
  ndx2 = 0
  # always add the smaller element first:
  while(ndx1 < len(items1) and ndx2 < len(items2)):
    #print("items1 data:",items1[ndx1],"items2
data:",items2[ndx2])

    data1 = items1[ndx1]
    data2 = items2[ndx2]
    if(data1 < data2):
      #print("adding data1:",data1)
      items3 = np.append(items3,data1)
      ndx1 += 1
    else:
      #print("adding data2:",data2)
      items3 = np.append(items3,data2)
      ndx2 += 1

  # append any remaining elements of items1:
  while(ndx1 < len(items1)):
      #print("MORE items1:",items1[ndx1])
      items3 = np.append(items3,data1)
      ndx1 += 1
```

```
    # append any remaining elements of items2:
    while(ndx2 < len(items2)):
        #print("MORE items2:",items2[ndx2])
        items3 = np.append(items3,data2)
        ndx2 += 1
    return items3

# display the merged list:
items3 = merge_items()
print("items1:",items1)
print("items2:",items2)
print("items3:",items3)
```

Listing 4.8 initializes the `NumPy` arrays `items1` and `items2`, followed by the function `merge_items()` that creates an empty `NumPy` array `items3` and the scalar variables `ndx1` and `ndx2` that keep track of the current index position in `items1` and `items2`, respectively.

The key idea is to compare the value of `items1[ndx1]` with the value of `items2[ndx2]`. If the smaller value is `items1[ndx1]`, then this value is appended to `items3` and `ndx1` is incremented. Otherwise, `items2[ndx2]` is appended to `items3` and `ndx2` is incremented.

The second part of the function `merge_items()` contains a loop that appends any remaining items in `items1` to `items3`, followed by another loop that appends any remaining items in `items2` to `items3`. The final portion of Listing 4.8 invokes the `merge_items()` function and then displays the contents of `items1`, `items2`, and `items3`.

There are several points to keep in mind regarding the code in Listing 4.8. First, the initial loop in `merge_items()` iterates through both `items1` and `items2` until the final element is reached in one of these two arrays: consequently, only one of these two arrays can be nonempty (and possibly both are empty), which means that only the second loop or the third loop is executed, but not both.

Second, this algorithm will only work if the elements in arrays `items1` and `items2` are sorted: to convince yourself that this is true, change the elements in `items1` (or in `items2`) so that they are no longer sorted and you will see that the output is incorrect. Third, this algorithm populates the array `items3` with the sorted list of values; later you will see an example of a merge sort that does not require the array `items3`. Now launch the code in Listing 4.8 and you will see the following output:

```
items1: [ 20  30  50 300]
items2: [ 80 100 200]
items3: [ 20  30  50  80 100 200 300]
```

Merge Sort Without a Third Array

Listing 4.9 displays the contents of the `merge_sort2.py` that illustrates how to perform a merge sort on two *sorted* arrays without using a third array.

LISTING 4.9: merge_sort2.py

```python
import numpy as np

items1 = np.array([20, 30, 50, 300, 0, 0, 0, 0])
items2 = np.array([80, 100, 200])

print("merge items2 into items1:")
print("INITIAL items1:",items1)
print("INITIAL items2:",items2)

def merge_arrays():
  ndx1 = 0
  ndx2 = 0
  last1 = 4 # do not count the 0 values

  # merge elements of items2 into items1:
  while(ndx2 < len(items2)):
    #print("items1 data:",items1[ndx1],"items2
data:",items2[ndx2])
    data1 = items1[ndx1]
    data2 = items2[ndx2]

    while(data1 < data2):
      prev1 = ndx1
      #print("incrementing ndx1:",ndx1)
      ndx1   += 1
      data1 = items1[ndx1]

      for idx3 in range(last1,ndx1,-1):
        #print("shift",items1[idx3],"to the right")
        items1[idx3] = items1[idx3-1]

      # insert data2 into items1:
      items1[ndx1] = data2
      ndx1   = 0
      ndx2   += 1
      last1 += 1
      #print("=> shifted items1:",items1)

merge_arrays()
print("UPDATED items1:",items1)
```

Although Listing 4.9 is an implementation of a merge sort algorithm, it differs from Listing 4.8 because a third array (such as items3 in Listing 4.8) is not required. As you can see, Listing 4.8 starts by initializing two Numpy arrays items1 and items2 with integer values, and then displays their contents.

However, there is a key difference: the right-most four elements of items1 are 0–these values will be replaced by the elements in items2, whose length is 3 (i.e., smaller than the number of available "zero" slots).

The next portion of Listing 4.9 defines the function `merge_arrays()` that starts by defining the scalar variables `ndx1` and `ndx2` that keep track of the current index position in `items1` and `items2`, respectively. The variable `last1` is initialized as 4, which is the right non-zero element in `items1`.

Listing 4.9 then defines a loop that iterates through the elements of `items2` in order to determine where to insert each of its elements in `items1`. Specifically, for each element `items2[ndx2]`, another loop determines the index position `ndx1` to insert `items2[ndx2]`. Note that before the insertion can be performed, the code shifts non-zero values to the right by one index position. As a result, an "open slot" becomes available for inserting `items2[ndx2]`. The final portion of Listing 4.9 invokes the function and then prints the contents of `items1`.

Keep in mind the following point: this code sample relies on the assumption that the right-most four values are 0 and that none of these values is a "legal" value in the array `items1`. However, the code sample in the next section removes this assumption. Now launch the code in Listing 4.9 and you will see the following output:

```
merge items2 into items1:
INITIAL items1: [ 20   30   50 300    0    0    0    0]
INITIAL items2: [ 80 100 200]
UPDATED items1: [ 20   30   50   80 100 200 300    0]
```

Merge Sort: Shift Elements From End of Lists

In this scenario we assume that matrix A has enough uninitialized elements at the end of the matrix in order to accommodate all the values of matrix B, as shown here:

```
        A              B              A
    +-----+        +-----+        +-----+
    |  20 |        |  50 |        |  20 |    A
    |  80 |        |  70 |        |  50 |    B
    | 200 |    +   | 100 |    =   |  70 |    B
    | 300 |        +-----+        |  80 |    A
    | 500 |                       | 100 |    B
    | xxx |                       | 200 |    A
    | xxx |                       | 300 |    A
    | xxx |                       | 500 |    A
    +-----+                       +-----+
```

Listing 4.10 displays the contents of the `merge_sort3.py` that illustrates how to perform a merge sort on two *sorted* arrays without using a third array.

LISTING 4.10: merge_sort3.py

```
import numpy as np

items1 = np.array([20, 30, 50, 300])
```

```
items2 = np.array([80, 100, 200])
last1 = len(items1)
last2 = len(items1)

print("=> merge items2 into items1 <=")
print("INITIAL items1:",items1)
print("INITIAL items2:",items2)

# append None to items1 for "empty slots":
for i in range(0,len(items2)):
  items1 = np.append(items1,None)
#print("AFTER  items1:",items1)

len1 = len(items1)
len2 = len(items2)

# start from the end of items1 and items2
# and shift items to the end of items1
def merge_arrays(items1,items2,len1,len2):
  ndx1 = len1-1
  ndx2 = len2-1
  last1 = len(items1)-1
  last2 = len(items2)-1

  # merge elements of items2 into items1:
  while(ndx1 >=0 and ndx2 >=0):
    #print("ndx1:",ndx1, "ndx2:",ndx2)
    data1 = items1[ndx1]
    data2 = items2[ndx2]
    #print("Bitems1 data:",data1,"ndx1:",ndx1)
    #print("Bitems2 data:",data2,"ndx2:",ndx2)

    if(data1 > data2):
      items1[last1] = data1
      ndx1 -= 1
      last1 -= 1
    else:
      items1[last1] = data2
      ndx2 -= 1
      last1 -= 1
    #print("Citems1:",items1)
    #print("Citems2:",items2)

merge_arrays(items1,items2,last1,len2)
print("MERGED items1:",items1)
```

The code for Listing 4.10 does not require an inner loop, and therefore Listing 4.10 is simpler than Listing 4.9. The code starts with two sorted arrays items1 and items2, after which items1 is padded with the value None so that it can accommodate the integers from items2. Launch the code in Listing 4.10 and you will see the following output:

```
merge items2 into items1:

=> merge items2 into items1 <=
INITIAL items1: [ 20  30  50 300]
```

```
INITIAL items2: [ 80 100 200]
MERGED items1: [20 30 50 80 100 200 300]
```

HOW DOES QUICK SORT WORK?

The *quick sort* algorithm uses a divide-and-conquer approach to sort an array of elements. The key idea involves selecting an item in a given list as the *pivot* item (which can be any item in the list) that is used for partitioning the given list into two sublists and then recursively sorting the two sublists.

Due to the recursive nature of this algorithm, each recursive invocation results in smaller sublists. Therefore, the sublists eventually reach the base cases where the sublists have either 0 or 1 elements (which are obviously sorted).

Another key point: one sublist contains values that are less than the pivot item, and the other sublist contains values that are greater than the pivot item. In the ideal case, both sublists have approximately the same length. This results in a binary-like splitting of the sublists, which involves log N invocations of the quick sort algorithm, where N is the number of elements in the list.

There are several points to keep in mind regarding the quick sort algorithm. First, the case in which the two sublists are approximately the same length is the more efficient case. However, this case involves a prior knowledge of the data distribution in the given list in order to achieve optimality.

Second, if the list contains values that are close to randomly distributed, in which case the first value or the last value are common choices for the pivot item. Third, quick sort has its worst performance when the values in a list are already sorted. In this scenario, select the pivot item in one of the following ways:

- Select the middle item in the list.
- Select the median of the first, middle, and last items in the list.

QUICK SORT CODE SAMPLE

Listing 4.11 displays the contents of the Python file quick_sort.py that illustrates how to perform a quick sort on an array of numbers.

LISTING 4.11: quick_sort.py

```python
def partition(start, end, array):
  # Initializing pivot's index to start
  pivot_index = start
  pivot = array[pivot_index]

  # This loop runs till start pointer crosses
  # end pointer, and when it does we swap the
  # pivot with element on end pointer
  while start < end:
    # Increment the start pointer till it finds an
    # element greater than  pivot
```

```
    while start < len(array) and array[start] <= pivot:
      start += 1

    # Decrement the end pointer till it finds an
    # element less than pivot
    while array[end] > pivot:
      end -= 1

    # If start and end have not crossed each other,
    # swap the numbers on start and end
    if(start < end):
      array[start], array[end] = array[end], array[start]

  # Swap pivot element with element on end pointer.
  # This puts pivot on its correct sorted place.
  array[end], array[pivot_index] = array[pivot_index],
array[end]

  # Returning end pointer to divide the array into 2
  return end

def quick_sort(start, end, array):
  if (start < end):
    # p is partitioning index, array[p] is at right place
    p = partition(start, end, array)

    # Sort elements before partition and after partition
    quick_sort(start, p - 1, array)
    quick_sort(p + 1, end, array)

array = [ 10, 7, 8, 9, 1, 5 ]
quick_sort(0, len(array) - 1, array)

print(f'Sorted array: {array}')
```

Listing 4.11 starts with the definition of the partition() function that takes three parameters start, end, and array, which represent the left index, the right index, and an array variable, respectively.

The next portion of Listing 4.11 is an outer loop that executes while the value of the start variable is less than the end variable. During each iteration of the outer loop, another while loop executes and increment start by 1 as long as the value of start is less than the length of array *and* the value of array[start] is less than or equal to the pivot value.

Next, another while loop decrements the value of the variable end as long as the value of array[end] is greater than the value of the end value. Thus, start moves in a left-to-right fashion through the elements of array, whereas end moves in a right-to-left fashion through the elements of array.

The next snippet of conditional logic checks if start is less than end, and if so, then array "swaps" the values in index start and index end, as shown here:

```
array[start], array[end] = array[end], array[start]
```

After the outer loop has completed execution, the next portion of the `par-tition()` function swaps the values of the index position `end` and the index position `pivot_index`, as shown here:

```
array[end], array[pivot_index] = array[pivot_index], array[end]
```

The final code snippet in the `partition()` function returns the value of the `end` variable.

The next portion of Listing 4.11 is the `quick_sort()` function that has the same parameters as the `partition()` function, along with conditional logic checks if `start` is less than `end`. If the latter is true, then the variable p is initialized with the result of invoking the `partition()` function, after which the `quick_sort()` function is recursively invoked twice, as shown here:

```
quick_sort(start, p - 1, array)
quick_sort(p + 1, end, array)
```

The final portion of Listing 4.11 initializes the variable `array` with a list of integers, invokes the `quick_sort()` function, and then displays the sorted array. Now launch the code in Listing 4.11 and you will see the following output:

```
Sorted array: [1, 5, 7, 8, 9, 10]
```

SHELLSORT

The Shellsort (by Donald L. Shell) is similar to the bubble sort: both algorithms compare pairs of elements and then swap them if they are out of order. However, unlike bubble sort, shell sort does not compare adjacent elements until the last iteration. Instead, it first compares elements that are widely separated, shrinking the size of the `gap` with each pass. In this way, it deals with elements that are significantly out of place early on, reducing the amount of work that later passes must do.

Listing 4.12 displays the contents of the `shell_sort.py` that illustrates how to implement the shell sort algorithm in `Python`.

LISTING 4.12: shell_sort.py

```python
def shell_sort(arr, num):
  gap = int(num/2)
  while(gap > 0):
    #for i in range(gap,num-1):
    for i in range(gap,num):
      j = i-gap
      while(j >= 0 and arr[j] > arr[j+gap]):
        # swap arr[j] and arr[j+gap]
        temp = arr[j]
        arr[j] = arr[j+gap]
        arr[j+gap] = temp
```

```
        j = j-gap
    gap = int(gap/2)
   return arr


num = 6
arr = [50,20,80,-100,500,200]
print("Original:",arr)
result = shell_sort(arr,num)
print("Sorted:  ",result)
```

Listing 4.12 defines the function `shell_sort()` whose two parameters are a list `arr` of numbers and an integer `num` that equals the length of the list. Next, the variable `gap` is initialized as `num/2`, followed by a `while` loop that executes as long as `num` is greater than 0.

Inside the `while` loop is a `for` loop that iterates through the integers from `gap` to `num`, which is initially from 3 to 6, respectively. The variable `j` is initialized as `i-gap`, which means that the initial value of `j` is 0.

The next `while` loop is the key portion of the code: notice that when a number on the left-side of `gap` is larger than a corresponding value on the right-side of `gap` is detected, this pair of numbers is "swapped" so that the numbers on the right-side of `gap` will all be greater than the numbers on the left-side of `gap`.

When the `while` loop has completed and its "parent" `for` loop has also completed, `gap` is replaced with `gap/2`, and the preceding process is repeated. When `gap` is assigned the value 0, the function returns the list `arr` that contains the reshuffled numbers that are ordered from smallest to largest. Launch the code in Listing 4.12 and you will see the following output:

```
Original: [50, 20, 80, -100, 500, 200]
Sorted:   [-100, 20, 50, 80, 200, 500]
```

SUMMARY

This chapter started with search algorithms, such as linear search and binary search (iterative and recursive). Next, you learned about the well-known bubble sort, selection sort, and insertion sort. You also saw how to perform the merge sort which can be performed in multiple ways. Finally, you learned about the quick sort and the shell sort.

## CHAPTER 5

LINKED LISTS

T his chapter contains an introduction to various types of linked list data structures, such as singly linked lists, doubly linked lists, and circular lists. These data structures are dynamically created, and therefore you do not need to know the number of elements in advance, which is an advantage of linked lists over linear lists (described later).

The first part of this chapter introduces you to singly linked lists, followed by examples of performing various operations on singly linked lists, such as creating and displaying the contents of linked lists, as well as updating nodes and deleting nodes in a singly linked list.

The second part of this chapter introduces you to doubly linked lists, followed by examples of performing various operations on doubly linked lists, which are the counterpart to the code samples for singly linked lists.

One important point to keep in mind for this chapter (as well as the other chapters) in this book: the code samples provide a solution that prefers clarity over optimization. In addition, many code samples contain "commented out" `print()` statements.

If the code samples confuse you, uncomment the `print()` statements in order to trace the execution path of the code: doing so can make the code easier to follow and also save you a lot of time. Indeed, after you have read each code sample and you fully understand the code, try to optimize the code, or perhaps use a different algorithm, which will enhance your problem solving skills as well as increase your confidence level.

TYPES OF DATA STRUCTURES

This section introduces you to the concept of linear data structures (stacks and queues) and nonlinear data structures (trees and graphs). This chapter and Chapter 6 are devoted to singly linked lists and doubly linked lists. This chapter contains the theoretical aspects of linked lists, whereas Chapter 6 contains tasks for which those data structures are well-suited (e.g., inserting new elements or finding existing elements).

Linear Data Structures

Linear data structures are data structures whose elements occur in sequential memory locations or are logically connected, such as stacks and queues. As you will see in Chapter 7, stacks are last-in-first-out (LIFO) data structures, which means that the last element inserted is the first element removed. Moreover, operations are performed from one end of the stack. A real-life counterpart to a stack is an elevator that has a single entrance that is also the exit.

In Chapter 7, you will learn about queues, which are first-in-first-out (FIFO) data structures, which means that the first element inserted is the first element removed. By contrast with a stack, insert operations are performed at the so-called "front" of the queue and delete operations are performed at the "rear" of the queue. A real-life counterpart to a queue is a line of people waiting to purchase tickets to a movie.

Nonlinear Data Structures

Nonlinear data structures are data structures whose elements are not sequential, such as trees and graphs.

A *tree* has a single node called the root node that has no parent node, whereas every other node has exactly one parent node. Two nodes are related if there is edge connecting the two nodes. In fact, the nodes in a tree have a hierarchical relationship.

A *graph* is a generalization of a tree, and nodes in a graph can have multiple parent nodes. Moreover, a graph does not have a root node: instead, a graph can have a "source" and a "sink" that are somewhat analogous to a "start" node and an "end" node. This designation appears in graphs that represent transport networks in which edges can have weights assigned to them. Think of trucks that transport food or other commodities from a warehouse (the "source") and have multiple routes to reach their destination (the "sink").

DATA STRUCTURES AND OPERATIONS

In this book, the data structure for singly linked lists is a custom `Node` class that consists of the following:

1. a key/value pair
2. a pointer to the next node

The data structure for doubly linked lists is a custom `Node` class that consists of the following:

1. a key/value pair
2. a pointer to next node
3. a pointer to the previous node

The simplest data structure for a stack is a `Python` list, which involves the following:

a LIFO (last-in-first-out) structure
a set of values (can be any type)
a method to check if the stack is empty or full
a method to insert a new element
a method to remove the "top" element

The simplest data structure for a queue is a `Python` list, which involves the following:

a FIFO (first-in-first-out) structure
a set of values (can be any type)
a method to check if the queue is empty
a method to insert a new element
a method to remove the first/front element

Note that Python provides a built-in Queue class that defines all the required methods for you. In addition, you could use a NumPy array to implement a stack or a queue.

The data structure for trees is a custom `Node` class that consists of the following:

1. a key/value pair
2. a pointer to a left (child) node
3. a pointer to a right (child) node

The data structure for a hash table is a `Python` dictionary. The preceding structures are based on simplicity; however, please keep in mind that other structures are possible as well. For example, you can use an array to implement a list, a stack, a queue, and even a tree.

Operations on Data Structures

In this chapter and the next, the operations on these data structures usually involve the following:

1. insert (which includes append)
2. delete
3. search
4. update (an existing element)

5. check for empty structure
6. check for full structure (queues and stacks)

WHAT ARE SINGLY LINKED LISTS?

Although arrays are useful data structures, one of their disadvantages is that the size or the number of elements in an array must be known in advance. One alternative that does not have this disadvantage is a "linked list," which is the topic of this chapter.

A singly linked list is a collection of data elements, which are commonly called nodes. Nodes contain two things:

1. a value that is stored in the node, and
2. the location of the next node (called its successor)

By way of analogy, think of a conga line of dances. Each dancer is a node, and the dancer places his or her hands on the hips of another dancer: the latter dancer is the location of the next node.

Unlike arrays, linked lists are dynamically created on an as-needed basis. Moreover, the preceding analogy makes the following point clear: there is a "last" node that does not have a next node. Thus, the last node in a list has None as its successor (i.e., next node).

In general, a node in a singly linked list can be one of the following three types:

1. the "root" node
2. an intermediate node
3. the last node (no next element)

Of course, when a linked list is empty, then the nodes in the preceding list are all None.

Trade-Offs for Linked Lists

Every data structure has trade-offs (i.e., advantages and disadvantages), and you need to consider these trade-offs before you select a data structure for your data.

Advantages of Linked Lists:

- Linked lists are dynamic data structures.
- There is efficient memory utilization.
- Memory is allocated whenever it is required.
- It is easy to deallocate memory when it is no longer needed.
- Insertion and deletions are easier and efficient.

- The number of elements in a list is not required in advance.
- No "shifting over" is required during operations.
- Elements of linked lists can be primary data type or user-defined data types.

Disadvantages of Linked Lists:

- Elements must be accessed sequentially from the first node (no random access).
- Binary search with linked lists is not possible.
- It is more difficult to sort a linked list.
- Accessing an element requires traversing (on average) half the list.
- It is more complex to create a linked list than an array.
- Extra memory is required for a pointer for every element in the list.

Thus, arrays work better when the number of elements is known in advance and there are no insertions or deletions (only updates), whereas linked lists are more efficient when insertions or deletions are allowed, neither of which requires knowing the number of data elements in advance.

The preceding trade-offs and observations apply to singly linked lists, circular lists, and doubly linked lists. Note that inserting a new node and deleting an existing node doubly linked lists are slightly more complicated than the same operations with singly linked lists.

The elements of linked lists are dynamically constructed, and unlike arrays, the elements of a linked list can be stored in noncontiguous memory locations: the value of the next field provides the location of the next node in the linked list (except for the LAST node that has None as its next node).

SINGLY LINKED LISTS: CREATE AND APPEND OPERATIONS

Linked lists support several operations, including insert (add a new node), delete (an existing node), update (change the value of an existing node), and traverse (list all the nodes). The following subsections discuss the preceding operations in more detail.

A Node Class for Singly Linked Lists

Listing 5.1 displays the contents of the Python file SLNode.py that illustrates how to define a simple Python class that represents a node in a singly linked list.

LISTING 5.1: SLNode.py

```
class SLNode:
   def __init__( self, data ):
      self.data = data
      self.next = None
```

```
node1 = SLNode("Jane")
node2 = SLNode("Dave")
node3 = SLNode("Stan")
node4 = SLNode("Alex")

print("node1.data:",node1.data)
print("node2.data:",node2.data)
print("node3.data:",node3.data)
print("node4.data:",node4.data)
```

Listing 5.1 is straightforward: it defines the Python class SLNode, followed by four instances of the SLNode class. The last portion of Listing 5.1 displays the contents of the four nodes. Now launch the following command from the command line and you will see the following output:

```
node1.data:  Jane
node2.data:  Dave
node3.data:  Stan
node4.data:  Alex
```

Appending a Node in a Linked List

When you create a linked list, you must *always* check if the root node is empty: if so, then you create a root node, otherwise you append the new node to the last node. Let's translate the preceding sentence into pseudocode that describes how to add a new element to a linked list:

```
Let ROOT be the root node (initially NULL) of the linked list
Let LAST be the last node (initially NULL) of the linked list
Let NEW be a new node and let NEW->next = NULL

# decide where to insert the new node:
if (ROOT == NULL)
{
    ROOT = NEW;
    LAST = NEW;
}
else
{
    LAST->next = NEW;
    LAST = NEW;
}
```

The last node in a linked list points to a NULL element.

Python Code for Appending a Node

Listing 5.2 displays the contents of the Python file append_slnode.py that illustrates a better way to create a linked list and append nodes to that list.

LISTING 5.2: append_slnode.py

```
import numpy as np

class SLNode:
```

```
    def __init__(self, data):
      self.data = data
      self.next = None

# a standalone function that is not part of SLNode
def append_node(ROOT, LAST, item):
  if(ROOT == None):
    ROOT = SLNode(item)
    #print("1Node:", ROOT.data)
  else:
    if(ROOT.next == None):
      NEWNODE = SLNode(item)
      LAST = NEWNODE
      ROOT.next = LAST
      #print("2Node:", NEWNODE.data)
    else:
      NEWNODE = SLNode(item)
      LAST.next = NEWNODE
      LAST = NEWNODE
      #print("3Node:", NEWNODE.data)

  return ROOT, LAST

ROOT = None
LAST = None

# append items to list:
items = np.array(["Stan", "Steve", "Sally", "Alex"])
for item in items:
  ROOT, LAST = append_node(ROOT, LAST, item)

# display items in list:
CURR = ROOT
while(CURR != None):
  print("Node:", CURR.data)
  CURR = CURR.next
```

Listing 5.2 defines a Node class as before, followed by the Python function append_node() that contains the logic for initializing a singly linked list and also for appending nodes to that list.

Now launch the code in Listing 5.2 from the command line and you will see the following output:

```
Node: Stan
Node: Steve
Node: Sally
Node: Alex
```

SINGLY LINKED LISTS: FINDING A NODE

The previous section showed you how to create a linked list by appending items to a singly linked list, whereas this section shows you how to find a node and how to insert a node in a singly linked list.

Listing 5.3 displays the contents of the Python file find_slnode.py that illustrates how to find a node in a linked list.

LISTING 5.3: find_slnode.py

```
class SLNode:
  def __init__(self, data):
    self.data = data
    self.next = None

def find_item(ROOT,item):
  found = False
  CURR = ROOT
  print("=> Search for:",item)
  while (CURR != None):
    print("Checking:",CURR.data)
    if(CURR.data == item):
      print("=> Found",item)
      found = True
      break;
    else:
      CURR = CURR.next

  if(found == False):
    print("*",item,"not found *")
  print("------------------\n")

ROOT  = None
LAST  = None

node1 = SLNode("Jane")
ROOT  = node1

node2 = SLNode("Dave")
LAST = node2
ROOT.next = LAST

node3 = SLNode("Stan")
LAST.next = node3
LAST  = node3

node4 = SLNode("Alex")
LAST.next = node4
LAST  = node4

items = np.array(["Stan", "Steve", "Sally", "Alex"])

for item in items:
  find_item(ROOT,item)
```

Listing 5.3 is straightforward: it initializes the variables msg and num with the specified values. Now launch the code in Listing 5.3 from the command line and you will see the following output:

```
=> Search for: Stan
Checking: Jane
Checking: Dave
Checking: Stan
=> Found Stan
------------------

=> Search for: Steve
Checking: Jane
Checking: Dave
Checking: Stan
Checking: Alex
* Steve not found *
------------------

=> Search for: Sally
Checking: Jane
Checking: Dave
Checking: Stan
Checking: Alex
* Sally not found *
------------------

=> Search for: Alex
Checking: Jane
Checking: Dave
Checking: Stan
Checking: Alex
=> Found Alex
```

Listing 5.3 is efficient for a small number of nodes. For linked lists that contain a larger number of nodes, we need a scalable way to construct a list of nodes, which is discussed in the next section.

Listing 5.4 displays the contents of the Python file find_slnode2.py that illustrates how to find a node in a singly linked list.

LISTING 5.4: find_slnode2.py (Method #2)

```python
import numpy as np

class SLNode:
  def __init__(self, data):
    self.data = data
    self.next = None

def append_node(ROOT, LAST, item):
  if(ROOT == None):
    ROOT = SLNode(item)
    #print("1Node:", ROOT.data)
  else:
    if(ROOT.next == None):
      NEWNODE = SLNode(item)
      LAST = NEWNODE
```

```
      ROOT.next = LAST
      #print("2Node:", NEWNODE.data)
    else:
      NEWNODE = SLNode(item)
      LAST.next = NEWNODE
      LAST = NEWNODE
      #print("3Node:", NEWNODE.data)

  return ROOT, LAST

def find_item(ROOT,item):
  found = False
  CURR = ROOT
  print("=> Search for:",item)
  while (CURR != None):
    print("Checking:",CURR.data)
    if(CURR.data == item):
      print("=> Found",item)
      found = True
      break;
    else:
      CURR = CURR.next

  if(found == False):
    print("*",item,"not found *")
  print("-------------------\n")

ROOT  = None
LAST  = None

items = np.array(["Stan", "Steve", "Sally", "Alex"])
for item in items:
  ROOT, LAST = append_node(ROOT, LAST, item)

for item in items:
  find_item(ROOT,item)
```

Listing 5.4 starts with the definition of the Python class SLNode, followed by the append() function that has ROOT, LAST, item as parameters. If ROOT is empty, then ROOT is assigned the SLNode instance that has item as its data value. If ROOT.next is empty, then ROOT.next is assigned the SLNode instance that has item as its data value. However, if ROOT.next is *not* None, then LAST.next assigned the SLNode instance that has item as its data value.

The next portion of Listing 5.4 defines the find_item() function that takes ROOT and item as parameters. Next, the variable CURR is initialized as ROOT, followed by a loop that iterates through the list of nodes. During each iteration, the value of CURR.item is compared with item: if they are equal, then the variable found is set equal to True and we exit the loop. Otherwise, if CURR.item never equals item, then the item will not be found and the variable found will maintain its initial value of False. The final portion of this function returns the value True or False depending on whether or not the value of found is True or False, respectively.

The next portion of Listing 5.4 initializes the variable `items` as a `NumPy` array of strings, followed by a loop that iterates through the values in `items`. During each iteration, the function `append_node()` is invoked in order to append a new node to a singly linked list. The final portion of Listing 5.4 also iterates through the elements in the variable `items` and invokes the `find_element()` function to search for each element (which obviously will be found in every case). Now launch the code in Listing 5.4 from the command line and you will see the following output:

```
node data: Jane
=> Search for: Stan
Checking: Stan
=> Found Stan
-------------------

=> Search for: Steve
Checking: Stan
Checking: Steve
=> Found Steve
-------------------

=> Search for: Sally
Checking: Stan
Checking: Steve
Checking: Sally
=> Found Sally
-------------------

=> Search for: Alex
Checking: Stan
Checking: Steve
Checking: Sally
Checking: Alex
=> Found Alex
-------------------
```

SINGLY LINKED LISTS: UPDATE AND DELETE OPERATIONS

In the previous section you saw how to find a node in a singly linked list. In this section you will learn how to update a node in a linked list as also how to delete a node in a linked list.

Updating a Node in a Singly Linked List

The following pseudocode explains how to search for an element, and update its contents if the element is present in a linked list.

```
CURR = ROOT
Found = False
```

```
OLDDATA = "something old";
NEWDATA = "something new";

If (ROOT == NULL)
{
   print("* EMPTY LIST *");
}

while (CURR != NULL)
{
   if(CURR->data = OLDDATA)
   {
      print("found node with value",OLDDATA);
      CURR->data = NEWDATA;
   }

   if(Found == True) { break; }

   PREV = CURR;
   CURR = CURR->next;
}
```

Python Code to Update a Node

Listing 5.5 displays the contents of the Python file update_slnode.py that illustrates how to update a node in a linked list.

LISTING 5.5: update_slnode.py

```python
import numpy as np

class SLNode:
  def __init__(self, data):
    self.data = data
    self.next = None

def append_node(ROOT, LAST, item):
  if(ROOT == None):
    ROOT = SLNode(item)
    #print("1Node:", ROOT.data)
  else:
    if(ROOT.next == None):
      NEWNODE = SLNode(item)
      LAST = NEWNODE
      ROOT.next = LAST
      #print("2Node:", NEWNODE.data)
    else:
      NEWNODE = SLNode(item)
      LAST.next = NEWNODE
      LAST = NEWNODE
      #print("3Node:", NEWNODE.data)

  return ROOT, LAST
```

```
ROOT  = None
LAST  = None

# append items to list:
items = np.array(["Stan", "Steve", "Sally", "Alex"])
for item in items:
  ROOT, LAST = append_node(ROOT, LAST, item)

# display items in list:
print("=> list items:")
CURR = ROOT
while(CURR != None):
  print("Node:", CURR.data)
  CURR = CURR.next
print()

# update item in list:
curr_val = "Alex"
new_val  = "Alexander"
found = False

CURR = ROOT
while(CURR != None):
  if(CURR.data == curr_val):
    print("Found:   ", CURR.data)
    CURR.data = new_val
    print("Updated:", CURR.data)
    found = True
    break
  else:
    CURR = CURR.next

if(found == False):
  print("* Item",curr_val,"not in list *")
```

Listing 5.5 defines a Node class as before, followed by the Python function append_node() that contains the logic for initializing a singly linked list and also for appending nodes to that list.

Now launch the code in Listing 5.5 from the command line and you will see the following output:

```
=> list items:
Node: Stan
Node: Steve
Node: Sally
Node: Alex

Found:   Alex
Updated: Alexander
```

DELETING A NODE IN A LINKED LIST: METHOD #1

The following pseudocode explains how to search for an element and then delete the element if it is present in a linked list:

```
CURR = ROOT
PREV = ROOT
item = <node-value>
Found = False

if (ROOT == NULL)
{
    print"* EMPTY LIST *");
}

while (CURR != NULL)
{
    if(CURR.data == item)
    {
        print("found node with value",item);

        Found = True
        if(CURR == ROOT)
        {
            ROOT = CURR.next // the list is now empty
        }
        else
        {
            PREV.next = CURR.next;
        }
    }

    if(found == True) { break; }

    PREV = CURR;
    CURR = CURR.next;
}

return ROOT
```

Now that you have seen the pseudocode, let's look at the code for deleting a node in a linked list, which is discussed in the next section.

PYTHON CODE FOR DELETING A NODE: METHOD #2

Listing 5.6 displays the contents of the Python file `delete_slnode.py` that illustrates how to delete a node in a linked list. This code sample is longer than the code samples in the previous sections because the code needs to distinguish between deleting the root node versus a nonroot node.

LISTING 5.6: delete_slnode.py

```python
import numpy as np

class SLNode:
    def __init__(self, data):
        self.data = data
        self.next = None
```

```python
def append_node(ROOT, LAST, item):
  if(ROOT == None):
    ROOT = SLNode(item)
    #print("1Node:", ROOT.data)
  else:
    if(ROOT.next == None):
      NEWNODE = SLNode(item)
      LAST = NEWNODE
      ROOT.next = LAST
      #print("2Node:", NEWNODE.data)
    else:
      NEWNODE = SLNode(item)
      LAST.next = NEWNODE
      LAST = NEWNODE
      #print("3Node:", NEWNODE.data)

  return ROOT, LAST

def delete_item(ROOT, item):
  PREV = ROOT
  CURR = ROOT
  found = False

  print("=> searching for item:",item)
  while (CURR != None):
    if(CURR.data == item):
      print("=> Found node with value",item)
      found = True

      if(CURR == ROOT):
        ROOT = CURR.next
        print("NEW ROOT")
      else:
        print("REMOVED NON-ROOT")
        PREV.next = CURR.next

    if(found == True):
      break

    PREV = CURR
    CURR = CURR.next

  if(found == False):
    print("* Item",item,"not in list *")

  return ROOT

def display_items(ROOT):
  print("=> list items:")
  CURR = ROOT
  while(CURR != None):
    print("Node:", CURR.data)
    CURR = CURR.next
  print()
```

```
ROOT  = None
LAST  = None

# append items to list:
items = np.array(["Stan", "Steve", "Sally",
"George","Alex"])
for item in items:
  ROOT, LAST = append_node(ROOT, LAST, item)

display_items(ROOT)

items2 = np.array(["Stan", "Alex", "Sally", "Steve",
"George"])
for item2 in items2:
  ROOT = delete_item(ROOT,item2)
  display_items(ROOT)
```

Listing 5.6 defines a Node class as before, followed by the Python function append_node() that contains the logic for initializing a singly linked list and also for appending nodes to that list. Now launch the code in Listing 5.6 from the command line and you will see the following output:

```
=> list items:
Node: Stan
Node: Steve
Node: Sally
Node: George
Node: Alex

=> searching for item: Stan
=> Found node with value Stan
NEW ROOT
=> list items:
Node: Steve
Node: Sally
Node: George
Node: Alex

=> searching for item: Alex
=> Found node with value Alex
REMOVED NON-ROOT
=> list items:
Node: Steve
Node: Sally
Node: George

=> searching for item: Sally
=> Found node with value Sally
REMOVED NON-ROOT
=> list items:
Node: Steve
Node: George
```

```
=> searching for item: Steve
=> Found node with value Steve
NEW ROOT
=> list items:
Node: George

=> searching for item: George
=> Found node with value George
NEW ROOT
=> list items:
```

CIRCULAR LINKED LISTS

The only structural difference between a singly linked list and a circular linked list is that the "last" node in a circular linked list has a "next" node equal to the initial (root) node. Operations on circular singly linked lists are the same as operations on singly linked lists, whereas operations on circular doubly linked lists are the same as operations on doubly linked lists. However, the algorithms for singly linked lists (and doubly linked lists) need to be modified in order to accommodate circular linked lists (and circular doubly linked lists).

Listing 5.7 displays the contents of the Python file circular_slnode.py that illustrates how to delete a node in a linked list. This code sample is longer than the code samples in the previous sections because the code needs to distinguish between deleting the root node versus a nonroot node.

LISTING 5.7: circular_slnode.py

```python
import numpy as np

class SLNode:
  def __init__(self, data):
    self.data = data
    self.next = None

def append_node(ROOT, LAST, item):
  if(ROOT == None):
    ROOT = SLNode(item)
    ROOT.next = ROOT
    LAST = ROOT
    print("1ROOT:", ROOT.data)
  else:
    if(ROOT.next == ROOT):
      NEWNODE = SLNode(item)
      LAST = NEWNODE
      ROOT.next = LAST
      LAST.next = ROOT
      print("2ROOT:", ROOT.data)
      print("2LAST:", LAST.data)
    else:
      NEWNODE = SLNode(item)
```

```
         NEWNODE.next = LAST.next
         LAST.next = NEWNODE
         LAST = NEWNODE
         print("3Node:", NEWNODE.data)

   return ROOT, LAST

ROOT  = None
LAST  = None

# insert items in circular linked list:
items = np.array(["Stan", "Steve", "Sally", "Alex"])
for item in items:
   ROOT, LAST = append_node(ROOT, LAST, item)
print()

# display items in list:
print("=> list items:")
CURR = ROOT
while(CURR != LAST):
   print("Node:", CURR.data)
   CURR = CURR.next

# print the last node as well:
print("Node:", LAST.data)
print()
```

Listing 5.7 defines a Node class as before, followed by the Python function append_node() that contains the logic for initializing a singly linked list and also for appending nodes to that list.

Now launch the code in Listing 5.7 from the command line and you will see the following output:

```
1ROOT: Stan
2ROOT: Stan
2LAST: Steve
3Node: Sally
3Node: Alex

=> list items:
Node: Stan
Node: Steve
Node: Sally
Node: Alex
```

PYTHON CODE FOR UPDATING A CIRCULAR LINKED LIST

As a reminder: the only structural difference between a singly linked list and a circular linked list is that the "last" node in the latter has a "next" node equal to the initial (root) node.

Listing 5.8 displays the contents of `circular_update_slnode.py` that illustrates how to delete a node in a linked list. This code sample is longer than the code samples in the previous sections because the code needs to distinguish between deleting the root node versus a nonroot node.

LISTING 5.8: circular_update_slnode.py

```
import numpy as np

class SLNode:
  def __init__(self, data):
    self.data = data
    self.next = None

def append_node(ROOT, LAST, item):
  if(ROOT == None):
    ROOT = SLNode(item)
    ROOT.next = ROOT
    LAST = ROOT
    print("1ROOT:", ROOT.data)
  else:
    if(ROOT.next == ROOT):
      NEWNODE = SLNode(item)
      LAST = NEWNODE
      ROOT.next = LAST
      LAST.next = ROOT
      print("2ROOT:", ROOT.data)
      print("2LAST:", LAST.data)
    else:
      NEWNODE = SLNode(item)
      NEWNODE.next = LAST.next
      LAST.next = NEWNODE
      LAST = NEWNODE
      print("3Node:", NEWNODE.data)

  return ROOT, LAST

# display items in list:
def display_items(ROOT):
  print("=> list items:")

  CURR = ROOT
  while(CURR != LAST):
    print("Node:", CURR.data)
    CURR = CURR.next

  # print the last node as well:
  print("Node:", LAST.data)
  print()

# update item in list:
def update_item(ROOT,curr_val,new_val):
```

```
    print("=> update list items:")

    found = False
    CURR = ROOT
    while(CURR != LAST):
      if(CURR.data == curr_val):
        print("Found data:", curr_val)
        CURR.data = new_val
        found = True
        break
      else:
        CURR = CURR.next

    # check the last node as well:
    if(found == False):
      if(LAST.data == curr_val):
        print("Found data in LAST:", curr_val)
        LAST.data = new_val
        found = True
    if(found == False):
      print("*",curr_val,"not found *")

    return ROOT, LAST

ROOT = None
LAST = None

# insert items in circular linked list:
items = np.array(["Stan", "Steve", "Sally", "Alex"])
for item in items:
  ROOT, LAST = append_node(ROOT, LAST, item)
print()

display_items(ROOT)

curr_val = "Alex"
new_val  = "Alexander"
ROOT, LAST = update_item(ROOT,curr_val,new_val)

display_items(ROOT)
```

Listing 5.8 defines a Node class as before, followed by the Python function append_node() that contains the logic for initializing a singly linked list and also for appending nodes to that list. Now launch the code in Listing 5.8 from the command line and you will see the following output:

```
1ROOT: Stan
1ROOT: Stan
2ROOT: Stan
2LAST: Steve
3Node: Sally
3Node: Alex
```

```
=> list items:
Node: Stan
Node: Steve
Node: Sally
Node: Alex

=> update list items:
Found data in LAST: Alex

=> list items:
Node: Stan
Node: Steve
Node: Sally
Node: Alexander
```

WORKING WITH DOUBLY LINKED LISTS (DLL)

A doubly linked list is a collection of data elements, which are commonly called nodes. Every node in a doubly linked list contains three items:

1. a value that is stored in the node, and
2. the location of the **next** node (called its successor)
3. The location of the **previous** node (called its predecessor)

Using the same "conga line" analogy as described in a previous section about singly linked lists, each person is touching the "next" node with one hand, and is touching the "previous" node with the other hand (not the best analogy, but you get the idea).

Operations on doubly linked lists are the same as the operations on singly linked lists; however, there are two pointers (the successor and the predecessor) to update instead of just one (the successor).

A Node Class for Doubly Linked Lists

Listing 5.9 displays the contents of the Python file DLNode.py that illustrates how to define a simple Python class that represents a single node in a linked list.

LISTING 5.9: DLNode.py

```
class DLNode:
  def __init__( self, data ):
    self.data = data
    self.next = None
    self.prev = None

node1 = DLNode("Jane")
node2 = DLNode("Dave")
node3 = DLNode("Stan")
node4 = DLNode("Alex")
```

```
print("node1.data:",node1.data)
print("node2.data:",node2.data)
print("node3.data:",node3.data)
print("node4.data:",node4.data)
```

Listing 5.9 is straightforward: it defines the `Python` class `DLNode` and then creates four such nodes. The final portion of Listing 5.9 displays the contents of the four nodes. Now launch the following command from the command line and you will see the following output:

```
node1.data:  Jane
node2.data:  Dave
node3.data:  Stan
node4.data:  Alex
```

Once again, a node in a doubly linked list can be one of the following three types:

1. the "root" node
2. an intermediate node
3. the last node (no next element)

Of course, when a linked list is empty, the nodes in the preceding list are all `None`.

APPENDING A NODE IN A DOUBLY LINKED LIST

When you create a linked list, you must *always* check if the root node is empty: if so, then you create a root node, otherwise you append the new node to the last node. Let's translate the preceding sentence into pseudocode that describes how to add a new element to a linked list:

```
Let ROOT be the root node (initially NULL) of the linked list
Let LAST be the last node (initially NULL) of the linked list
Let NEW be a new node with NEW->next = NULL and NEW->prev = NULL

# decide where to insert the new node:
if (ROOT == NULL)
{
   ROOT = NEW;
   ROOT->next = NULL;
   ROOT->prev = NULL;

   LAST = NEW;
   LAST->next = NULL;
   LAST->prev = NULL;
}
else
{
   NEW->prev = LAST;
   LAST->next = NEW;
   NEW->next = NULL;
   LAST = NEW;
}
```

The last node in a doubly linked list in Python always points to a None element, whereas pseudocode typically uses NULL or NIL as the counterpart to a None element.

Python Code for Appending a Node

Listing 5.10 displays the contents of the Python file append_dlnode.py that illustrates a scalable way to create a linked list and append nodes to that list.

LISTING 5.10: append_dlnode.py

```python
import numpy as np

class DLNode:
  def __init__(self, data):
    self.data = data
    self.prev = None
    self.next = None

def append_node(ROOT, LAST, item):
  if(ROOT == None):
    ROOT = DLNode(item)
    print("1Node:", ROOT.data)
  else:
    if(ROOT.next == None):
      NEWNODE = DLNode(item)
      NEWNODE.prev = ROOT
      LAST = NEWNODE

      ROOT.next = NEWNODE
      print("2Node:", NEWNODE.data)
    else:
      NEWNODE = DLNode(item)
      NEWNODE.prev = LAST

      LAST.next = NEWNODE
      LAST = NEWNODE
      #print("3Node:", NEWNODE.data)
  return ROOT, LAST

ROOT  = None
LAST  = None

# append items to list:
items = np.array(["Stan", "Steve", "Sally", "Alex"])
for item in items:
  ROOT, LAST = append_node(ROOT, LAST, item)

# display items in list:
CURR = ROOT
while(CURR != None):
  print("Node:", CURR.data)
  CURR = CURR.next
```

Listing 5.10 defines a Node class as before, followed by the Python function append_node() that contains the logic for initializing a singly linked list and also for appending nodes to that list. Now launch the code in Listing 5.10 from the command line and you will see the following output:

```
1ROOT: Stan
2ROOT: Stan
2LAST: Steve
3Node: Sally
3Node: Alex

=> list items:
Node: Stan
Node: Steve
Node: Sally
Node: Alex

=> update list items:
Found data in LAST: Alex

=> list items:
Node: Stan
Node: Steve
Node: Sally
Node: Alexander
```

Python Code for Inserting an Intermediate Node

Listing 5.11 displays the contents of the Python file new_inter_slnode. py that illustrates a scalable way to create a linked list and append nodes to that list.

LISTING 5.11: new_inter_slnode.py

```python
import numpy as np

class SLNode:
  def __init__(self, data):
    self.data = data
    self.next = None

# display items in list:
def display_list(ROOT):
  CURR = ROOT
  while(CURR != None):
    print("Node:", CURR.data)
    CURR = CURR.next

def append_node(ROOT, LAST, item):
  if(ROOT == None):
    ROOT = SLNode(item)
    #print("1Node:", ROOT.data)
  else:
```

```python
    if(ROOT.next == None):
      NEWNODE = SLNode(item)
      LAST = NEWNODE
      ROOT.next = LAST
      #print("2Node:", NEWNODE.data)
    else:
      NEWNODE = SLNode(item)
      LAST.next = NEWNODE
      LAST = NEWNODE
      #print("3Node:", NEWNODE.data)

  return ROOT, LAST

# insert intermediate item in list:
def insert_node(ROOT, new_val):
  CURR = ROOT
  PREV = CURR
  while(CURR != None):
    #print("Node:", CURR.data)
    if( CURR.data < new_val):
      if( CURR == ROOT):
        ROOT2 = SLNode(new_val)
        ROOT2.next = ROOT
        ROOT = ROOT2
        #print("new root:", ROOT.data)
        break
      elif( CURR.next == None):
        print("found last element:", CURR.data)
        PENULT = SLNode(new_val)
        PREV.next = PENULT
        PENULT.next = CURR
        break
      else:
        print("intermediate element:", CURR.data)
        INTERM = SLNode(new_val)
        INTERM.next = CURR
        PREV.next = INTERM
        break
    elif( CURR.next == None):
      # the new item is smaller than the last element:
      LAST = SLNode(new_val)
      CURR.next = LAST
      break

    PREV = CURR
    CURR = CURR.next
  return ROOT

ROOT  = None
LAST  = None

# append items to list:
items = np.array([200, 150, 100, 10])
for item in items:
  ROOT, LAST = append_node(ROOT, LAST, item)
```

```
#new_val = 250
#new_val = 175
#new_val = 50
new_val = 5

print("Initial List:")
display_list(ROOT)
print("=> Insert Value:",new_val)
ROOT = insert_node(ROOT, new_val)
print("Updated List:")
display_list(ROOT)
```

Listing 5.11 defines a `Node` class as before, followed by the `Python` function `append_node()` that contains the logic for initializing and appending nodes to a singly linked list.

The function `insert_node()` handles four cases that can arise when inserting a new node into a singly linked list:

- a new root node
- a new node in a list with only one node
- a new final node
- an intermediate node before the final node

The final portion of Listing 5.11 initializes the `Numpy` array with several integer values and displays the contents of the array. Next, the function `insert_node()` inserts a node with a new value. Notice that the code does not check if the new value is equal to an existing value: you can modify the code to handle this scenario. Now launch the code in Listing 5.11 from the command line and you will see the following output:

```
Initial List:
Node: 200
Node: 150
Node: 100
Node: 10
=> Insert Value: 5
Updated List:
Node: 200
Node: 150
Node: 100
Node: 10
Node: 5
```

SEARCHING AND UPDATING A NODE IN A DOUBLY LINKED LIST

The following pseudocode explains how to traverse the elements of a linked list, after which you will learn how to update the contents of a given node:

```
CURR = ROOT

while (CURR != NULL)
```

```
{
   print("contents:",CURR->data);
   CURR = CURR->next;
}

If (ROOT == NULL)
{
   print("* EMPTY LIST *");
}
```

Updating a Node in a Doubly Linked List

The following pseudocode explains how to search for an element: if the element is present in a linked list, then the code updates its contents:

```
CURR = ROOT
Found = False
OLDDATA = "something old";
NEWDATA = "something new";

If (ROOT == NULL)
{
   print("* EMPTY LIST *");
}

while (CURR != NULL)
{
   if(CURR->data = OLDDATA)
   {
      print("found node with value",OLDDATA);
      CURR->data = NEWDATA;
   }

   if(Found == True) { break; }

   PREV = CURR;
   CURR = CURR->next;
}
```

As you now know, some operations on doubly linked lists involve updating a single pointer, and other operations involve updating two pointers.

Python Code to Update a Node

Listing 5.12 displays the contents of the Python file update_dlnode.py that illustrates how to update a node in a linked list.

LISTING 5.12: update_dlnode.py

```
import numpy as np

class DLNode:
   def __init__(self, data):
      self.data = data
```

```python
      self.prev = None
      self.next = None

  # display items in list:
  def display_items(ROOT):
    print("=> items in list:")
    CURR = ROOT
    while(CURR != None):
      print("Node:", CURR.data)
      CURR = CURR.next
    print()


  def append_node(ROOT, LAST, item):
    if(ROOT == None):
      ROOT = DLNode(item)
      #print("1Node:", ROOT.data)
    else:
      if(ROOT.next == None):
        NEWNODE = DLNode(item)
        NEWNODE.prev = ROOT
        LAST = NEWNODE

        ROOT.next = NEWNODE
        #print("2Node:", NEWNODE.data)
      else:
        NEWNODE = DLNode(item)
        NEWNODE.prev = LAST

        LAST.next = NEWNODE
        LAST = NEWNODE
        #print("3Node:", NEWNODE.data)
    return ROOT, LAST

  # update list item:
  def update_item(ROOT,curr_val,new_val):
    found = False
    CURR = ROOT

    while(CURR != None):
      if(CURR.data == curr_val):
        print("Found data:",curr_val)
        CURR.data = new_val
        print("New value: ",new_val)
        found = True
        break
      else:
        CURR = CURR.next

ROOT  = None
LAST  = None


# create an array of items:
items = np.array(["Stan", "Steve", "Sally", "Alex"])
```

```
for item in items:
  ROOT, LAST = append_node(ROOT, LAST, item)

display_items(ROOT)

curr_val = "Alex"
new_val = "Alexander"
update_item(ROOT, curr_val, new_val)
display_items(ROOT)
```

Listing 5.12 defines a Node class as before, followed by the Python function append_node() that contains the logic for initializing a singly linked list and also for appending nodes to that list. Now launch the code in Listing 5.12 from the command line and you will see the following output:

```
Found:    Alex
=> items in list:
Node: Stan
Node: Steve
Node: Sally
Node: Alex

Found data: Alex
New value:  Alexander
=> items in list:
Node: Stan
Node: Steve
Node: Sally
Node: Alexander
```

DELETING A NODE IN A DOUBLY LINKED LIST

The following pseudocode explains how to search for an element, and also delete the element if it is present in a linked list:

```
CURR = ROOT
PREV = ROOT
ANODE = <a-node-to-delete>
Found = False

If (ROOT == NULL)
{
   print("* EMPTY LIST *");
}

while (CURR != NULL)
{
   if(CURR->data = ANODE->data)
   {
     print("found node with value",ANODE->data);

     Found = True
     if(CURR == ROOT)
     {
```

```
            ROOT = NULL; // the list is now empty
        }
        else
        {
            PREV->next = CURR->next;
        }
    }

    if(Found == True) { break; }

    PREV = CURR;
    CURR = CURR->next;
}
```

Python Code to Delete a Node

Listing 5.13 displays the contents of the Python file delete_dlnode.py that illustrates how to delete a node in a doubly linked list.

LISTING 5.13: delete_dlnode.py

```
import numpy as np

class DLNode:
  def __init__(self, data):
    self.data = data
    self.prev = None
    self.next = None

def append_node(ROOT, LAST, item):
  if(ROOT == None):
    ROOT = DLNode(item)
    print("1Node:", ROOT.data)
  else:
    if(ROOT.next == None):
      NEWNODE = DLNode(item)
      NEWNODE.prev = ROOT
      LAST = NEWNODE

      ROOT.next = NEWNODE
      print("2Node:", NEWNODE.data)
    else:
      NEWNODE = DLNode(item)
      NEWNODE.prev = LAST

      LAST.next = NEWNODE
      LAST = NEWNODE
      #print("3Node:", NEWNODE.data)
  return ROOT, LAST

def display_nodes(ROOT):
  CURR = ROOT
```

```
  while(CURR != None):
    print("Node value:",CURR.data)
    CURR = CURR.next


def delete_node(ROOT):
  CURR = ROOT
  while(CURR != None):
    if(CURR.data == item_val):
      print("Found:  ", CURR.data)

      # delete the current node:
      if(CURR == ROOT):
        print("2matching root:",ROOT.data)
        if(CURR.next == None):
          ROOT = None
        else:
          CURR.next.prev = ROOT
          ROOT = ROOT.next
          print("updated new root:",ROOT.data)
        break
      else:
        PREV = CURR.prev
        PREV.next = CURR.next
        if(CURR.next != None):
          CURR.next.prev = PREV

      found = True
      break
    else:
      CURR = CURR.next
  print("3returning root:",ROOT.data)
  return(ROOT)


ROOT  = None
LAST  = None

# append items to list:
items = np.array(["Stan", "Steve", "Sally", "Alex"])
for item in items:
  ROOT, LAST = append_node(ROOT, LAST, item)

# display items in list:
print("=> list items:")
display_nodes(ROOT)

# remove item from list:
item_val = "Alex"
found = False

# items = np.array(["Stan", "Steve", "Sally", "Alex"])

ROOT = delete_node(ROOT)
display_nodes(ROOT)
```

Listing 5.13 defines a `Node` class as before, followed by the `Python` function `append_node()` that contains the logic for initializing a doubly linked list and also for appending nodes to that list. Now launch the code in Listing 5.13 from the command line and you will see the following output:

```
1Node: Stan
2Node: Steve
=> list items:
Node value: Stan
Node value: Steve
Node value: Sally
Node value: Alex
Found:    Alex
3returning root: Stan
Node value: Stan
Node value: Steve
Node value: Sally
```

SUMMARY

This chapter started with a description of linked lists, along with their advantages and disadvantages. Then you learned how to perform several operations on singly linked lists such as append, insert, delete, and update.

Next, you learned about doubly linked lists, and how to perform the same operations on doubly linked lists that you performed on singly linked lists. You also saw how to work with circular lists in which the last element "points" to the first element in a list.

## CHAPTER 6

LINKED LISTS AND COMMON TASKS

The previous chapter introduced you to singly linked lists, doubly linked lists, and circular lists, and how to perform basic operations on those data structures. This chapter shows you how to perform a variety of tasks that involve more than the basic operations in the previous chapter.

The first part of this chapter contains code samples for displaying the first k nodes in a linked list as well as the last k nodes of a linked list. This section also shows you how to display the contents of a list in reverse order and how to remove duplicates.

The second part of this chapter contains code samples for concatenating two linked lists, merging two linked lists, and splitting a single linked list. In addition you will see how to remove the middle element in a list and how to determine whether or not a linked list contains a loop. The final code sample checks for palindromes in a linked list.

TASK: ADDING NUMBERS IN A LINKED LIST (1)

Listing 6.1 displays the contents of the Python file sum_slnodes.py that illustrates how to append a set of numbers in a linked list.

LISTING 6.1: sum_slnodes.py

```python
import numpy as np

class SLNode:
  def __init__(self, data):
    self.data = data
    self.next = None
```

```
def append_node(ROOT, LAST, item):
  if(ROOT == None):
    ROOT = SLNode(item)
    ROOT.next = ROOT
    LAST = ROOT
    print("1ROOT:", ROOT.data)
  else:
    if(ROOT.next == ROOT):
      NEWNODE = SLNode(item)
      LAST = NEWNODE
      ROOT.next = LAST
      LAST.next = ROOT
      print("2ROOT:", ROOT.data)
      print("2LAST:", LAST.data)
    else:
      NEWNODE = SLNode(item)
      NEWNODE.next = LAST.next
      LAST.next = NEWNODE
      LAST = NEWNODE
      print("3Node:", NEWNODE.data)

  return ROOT, LAST

ROOT  = None
LAST  = None

# append items to list:
items = np.array([1,2,3,4])
for item in items:
  ROOT, LAST = append_node(ROOT, LAST, item)

# compute the sum of the numbers:
sum = 0
CURR = ROOT
while(CURR != None):
  sum += CURR.data
  CURR = CURR.next

print("list of numbers:",items)
print("Sum of numbers: ",sum)
```

Listing 6.1 defines a Node class as before, followed by the Python function append_node() that contains the logic for initializing a singly linked list and also for appending nodes to that list.

Now launch the code in Listing 6.1 from the command line and you will see the following output:

```
list of numbers:  [1 2 3 4]
Sum of numbers:   10
```

TASK: RECONSTRUCTING NUMBERS IN A LINKED LIST (1)

Listing 6.2 displays the contents of the Python file sum_slnodes.py that illustrates how to add the numbers in a linked list.

LISTING 6.2: *sum_slnodes.py*

```python
import numpy as np

class SLNode:
  def __init__(self, data):
    self.data = data
    self.next = None

def append_node(ROOT, LAST, item):
  if(ROOT == None):
    ROOT = SLNode(item)
    ROOT.next = ROOT
    LAST = ROOT
    print("1ROOT:", ROOT.data)
  else:
    if(ROOT.next == ROOT):
      NEWNODE = SLNode(item)
      LAST = NEWNODE
      ROOT.next = LAST
      LAST.next = ROOT
      print("2ROOT:", ROOT.data)
      print("2LAST:", LAST.data)
    else:
      NEWNODE = SLNode(item)
      NEWNODE.next = LAST.next
      LAST.next = NEWNODE
      LAST = NEWNODE
      print("3Node:", NEWNODE.data)

  return ROOT, LAST

# append items to list:
items = np.array([1,2,3,4])
for item in items:
  ROOT, LAST = append_node(ROOT, LAST, item)

# note: [1,2,3,4] => 4,321
# compute the sum of the numbers:
sum  = 0
pow  = 0
base = 10

CURR = ROOT
while(CURR != None):
  term = CURR.data * (base**pow)
  sum += term
  pow += 1
  CURR = CURR.next

print("list of digits:  ",items)
print("Original number: ",sum)
```

Listing 6.2 defines a `Node` class as before, followed by the `Python` function `append_node()` that contains the logic for initializing a singly linked list and also for appending nodes to that list.

Now launch the code in Listing 6.2 from the command line and you will see the following output:

```
list of digits:   [1 2 3 4]
Original number:   4321
```

TASK: RECONSTRUCTING NUMBERS IN A LINKED LIST (2)

Listing 6.3 displays the contents of the `Python` file `sum_slnodes.py` that illustrates how to add the numbers in a linked list.

LISTING 6.3: sum_slnodes.py

```python
import numpy as np

class SLNode:
  def __init__(self, data):
    self.data = data
    self.next = None

def append_node(ROOT, LAST, item):
  if(ROOT == None):
    ROOT = SLNode(item)
    ROOT.next = ROOT
    LAST = ROOT
    print("1ROOT:", ROOT.data)
  else:
    if(ROOT.next == ROOT):
      NEWNODE = SLNode(item)
      LAST = NEWNODE
      ROOT.next = LAST
      LAST.next = ROOT
      print("2ROOT:", ROOT.data)
      print("2LAST:", LAST.data)
    else:
      NEWNODE = SLNode(item)
      NEWNODE.next = LAST.next
      LAST.next = NEWNODE
      LAST = NEWNODE
      print("3Node:", NEWNODE.data)

  return ROOT, LAST

def reverse_sum(node,sum,pow,base):
  if(node == None):
    print("1reverse: empty node")
    return sum

  if (node.next == None):
    print("2no child node value:",node.data)
```

```
      term = node.data * (base**pow)
      print("returning sum:",sum+term)
      return sum + term

  if (node.next != None):
    term = node.data * (base**pow)
    print("data:",node.data,"3finding next node...")
    return reverse_sum(node.next, sum+term, pow+1, base)
    print("4node value:",node.data,"sum:",sum,"pow:",pow)

ROOT  = None
LAST  = None

# append items to list:
items = np.array([1,2,3,4])
for item in items:
  ROOT, LAST = append_node(ROOT, LAST, item)

# note: [1,2,3,4] => 1,234
# compute the sum of the numbers:
sum  = 0
pow  = 0
base = 10

print("list of digits: ",items)
sum = reverse_sum(ROOT,sum,pow,base)
print("Reversed sum:",sum)
```

Listing 6.3 defines a Node class as before, followed by the Python function append_node() that contains the logic for initializing a singly linked list and also for appending nodes to that list.

Now launch the code in Listing 6.3 from the command line and you will see the following output:

```
list of digits:  [1 2 3 4]
data: 1 3finding next node...
data: 2 3finding next node...
data: 3 3finding next node...
2no child node value: 4
returning sum: 4321
Reversed sum: 4321
```

TASK: DISPLAY THE FIRST K NODES

Listing 6.4 displays the contents of the Python file first_k_nodes.py that illustrate how to find the first k nodes in a linked list.

LISTING 6.4: first_k_nodes.py

```
import numpy as np

class SLNode:
  def __init__(self, data):
```

```
      self.data = data
      self.next = None

def append_node(ROOT, LAST, item):
  if(ROOT == None):
    ROOT = SLNode(item)
    #print("1Node:", ROOT.data)
  else:
    if(ROOT.next == None):
      NEWNODE = SLNode(item)
      LAST = NEWNODE
      ROOT.next = LAST
      #print("2Node:", NEWNODE.data)
    else:
      NEWNODE = SLNode(item)
      LAST.next = NEWNODE
      LAST = NEWNODE
      #print("3Node:", NEWNODE.data)

  return ROOT, LAST

def first_k_nodes(ROOT, num):
  count = 0
  CURR = ROOT
  print("=> Display first",num,"nodes")
  while (CURR != None):
    count += 1
    print("Node",count,"data:",CURR.data)
    CURR = CURR.next

    if(count >= num):
      break

ROOT = None
LAST = None

# append items to list:
items = np.array(["Stan", "Steve", "Sally",
"Alex","George","Fred","Bob"])

for item in items:
  ROOT, LAST = append_node(ROOT, LAST, item)

print("initial list:")
print(items)
first_k_nodes(ROOT, 3)
```

Listing 6.4 defines a Node class as before, followed by the Python function append_node() that contains the logic for initializing a singly linked list and also for appending nodes to that list. Now launch the code in Listing 6.4 from the command line and you will see the following output:

```
initial list:
['Stan' 'Steve' 'Sally' 'Alex' 'George' 'Fred' 'Bob']
```

```
=> Display first 3 nodes
Node 1 data: Stan
Node 2 data: Steve
Node 3 data: Sally
```

TASK: DISPLAY THE LAST K NODES

Listing 6.5 displays the contents of the Python file last_k_nodes.py that illustrates how to find the last k nodes in a singly linked list.

LISTING 6.5: last_k_nodes.py

```python
import numpy as np

class SLNode:
  def __init__(self, data):
    self.data = data
    self.next = None

def append_node(ROOT, LAST, item):
  if(ROOT == None):
    ROOT = SLNode(item)
    #print("1Node:", ROOT.data)
  else:
    if(ROOT.next == None):
      NEWNODE = SLNode(item)
      LAST = NEWNODE
      ROOT.next = LAST
      #print("2Node:", NEWNODE.data)
    else:
      NEWNODE = SLNode(item)
      LAST.next = NEWNODE
      LAST = NEWNODE
      #print("3Node:", NEWNODE.data)

  return ROOT, LAST

def count_nodes(ROOT):
  count = 0
  CURR = ROOT
  while (CURR != None):
    count += 1
    CURR = CURR.next
  return count

def skip_nodes(ROOT,skip_count):
  count = 0
  CURR = ROOT
  while (CURR != None):
    count += 1
    CURR = CURR.next
    if(count >= skip_count):
      break
  return CURR
```

```
def last_k_nodes(ROOT, node_count, num):
  count = 0
  node_count = count_nodes(ROOT)
  START_NODE = skip_nodes(ROOT,node_count-num)

  CURR = START_NODE
  while (CURR != None):
    count += 1
    print("Node",count,"data:",CURR.data)
    CURR = CURR.next

    if(count >= num):
      break

ROOT  = None
LAST  = None

# append items to list:
items = np.array(["Stan", "Steve", "Sally",
"Alex","George","Fred","Bob"])

for item in items:
  ROOT, LAST = append_node(ROOT, LAST, item)

print("=> Initial list:")
print(items)
print()

list_length = count_nodes(ROOT)
node_count = 3
print("Last",node_count,"nodes:")
last_k_nodes(ROOT, list_length,node_count)
```

Listing 6.5 defines a Node class as before, followed by the Python function append_node() that contains the logic for initializing a singly linked list and then appending nodes to that list. There are three cases:

1. an empty list
2. a single-node list
3. a list with two or more nodes

Each of the three preceding cases is handled in the append_node() function. Now launch the code in Listing 6.5 from the command line and you will see the following output:

```
=> Initial list:
['Stan' 'Steve' 'Sally' 'Alex' 'George' 'Fred' 'Bob']

Last 3 nodes:
Node 1 data: George
Node 2 data: Fred
Node 3 data: Bob
```

DISPLAY A SINGLY LINKED LIST IN REVERSE ORDER VIA RECURSION

Listing 6.6 displays the contents of the `Python` file `reverse_sllist.py` that illustrates how to reverse the contents of a linked list.

LISTING 6.6: reverse_sllist.py

```python
import numpy as np

class SLNode:
  def __init__(self, data):
    self.data = data
    self.next = None

def append_node(ROOT, LAST, item):
  if(ROOT == None):
    ROOT = SLNode(item)
    #print("1Node:", ROOT.data)
  else:
    if(ROOT.next == None):
      NEWNODE = SLNode(item)
      LAST = NEWNODE
      ROOT.next = LAST
      #print("2Node:", NEWNODE.data)
    else:
      NEWNODE = SLNode(item)
      LAST.next = NEWNODE
      LAST = NEWNODE
      #print("3Node:", NEWNODE.data)

  return ROOT, LAST

def reverse_list(node, rev_list):
  if(node == None):
    print("1rev_list:",rev_list)
    return rev_list

  if (node.next == None):
    #print("2rev_list:",rev_list)
    return [node.data] + rev_list

  if (node.next != None):
    #print("3finding next node...")
    rev_list = [node.data] + rev_list
    #print("3rev_list:",rev_list)
    return reverse_list(node.next, rev_list)

ROOT  = None
LAST  = None

# append items to list:
items = np.array(["Stan", "Steve", "Sally", "Alex"])
```

```
for item in items:
  ROOT, LAST = append_node(ROOT, LAST, item)

# display items in list:
print("=> Original list items:")
CURR = ROOT
while(CURR != None):
  print("Node:", CURR.data)
  CURR = CURR.next
print()

print("=> Reversed list of items:")
rev_list = []
reversed = reverse_list(ROOT, rev_list)
print(reversed)
```

Listing 6.6 defines a Node class as before, followed by the Python function append_node() whose contents are the same as the corresponding code in Listing 6.5. The next portion of Listing 6.6 contains the logic for initializing a singly linked list and also for appending nodes to that list.

The next portion of Listing 6.6 defines the function reverse_list() that is invoked recursively in order to reverse the order of the elements in the parameter rev_list (which is a list of elements). Each time that the function reverse_list() is invoked, there are three possible cases. First, the parameter node might be empty, in which case rev_list is returned. The second possibility is that node does not have a successor node: in this case, the code invokes the reverse_list() function as shown here:

```
return reverse_list(node.next, rev_list)
```

As you can see, the *successor* of node becomes the new node to process in reverse_list().

The third possibility is that node and node.next are nonempty, in which case the following code snippet is executed:

```
return reverse_list(node.next, rev_list)
```

The next portion of Listing 6.6 initializes items as a NumPy array of strings, followed by a loop that constructs a linked list from the array items.

The final block of code displays the contents of the linked list, followed by an invocation of the reverse_list() function in order to reverse the elements in the constructed linked list:

```
reversed = reverse_list(ROOT, rev_list)
```

Now launch the code in Listing 6.6 from the command line and you will see the following output:

```
=> Original list items:
Node: Stan
Node: Steve
Node: Sally
Node: Alex

=> Reversed list of items:
['Alex', 'Sally', 'Steve', 'Stan']
```

TASK: REMOVE DUPLICATE NODES

Listing 6.7 displays the contents of the Python file remove_duplicates. py that illustrates how to remove duplicate nodes in a linked list.

LISTING 6.7: remove_duplicates.py

```
import numpy as np

class SLNode:
  def __init__(self, data):
    self.data = data
    self.next = None

def append_node(ROOT, LAST, item):
  if(ROOT == None):
    ROOT = SLNode(item)
    #print("1Node:", ROOT.data)
  else:
    if(ROOT.next == None):
      NEWNODE = SLNode(item)
      LAST = NEWNODE
      ROOT.next = LAST
      #print("2Node:", NEWNODE.data)
    else:
      NEWNODE = SLNode(item)
      LAST.next = NEWNODE
      LAST = NEWNODE
      #print("3Node:", NEWNODE.data)

  return ROOT, LAST

def delete_duplicates(ROOT):
  PREV = ROOT
  CURR = ROOT
  found = False

  print("=> searching for duplicates"w)
  duplicate = 0
  while (CURR != None):
    SEEK = CURR
    while (SEEK.next != None):
```

```
        if(SEEK.next.data == CURR.data):
          duplicate += 1
          print("=> Found duplicate node #",duplicate,"with
value:",item)
          SEEK.next = SEEK.next.next
        else:
          SEEK = SEEK.next
      CURR = CURR.next
    return ROOT

def display_items(ROOT):
  print("=> list items:")
  CURR = ROOT
  while(CURR != None):
    print("Node:", CURR.data)
    CURR = CURR.next
  print()

ROOT  = None
LAST  = None


# append items to list:
items = np.array(["Stan", "Steve", "Stan",
"George","Stan"])
for item in items:
  ROOT, LAST = append_node(ROOT, LAST, item)

display_items(ROOT)

items2 = np.array(["Stan", "Alex", "Sally", "Steve",
"George"])
for item2 in items2:
  ROOT = delete_duplicates(ROOT)
  display_items(ROOT)

items3 = np.array(["Stan", "Steve", "Stan",
"George","Stan"])
print("original:")
print(items3)

print("unique:")
print(display_items(ROOT))
```

Listing 6.7 defines a Node class as before, followed by the Python function append_node() that contains the logic for initializing a singly linked list and also for appending nodes to that list.

The new function in Listing 6.7 is delete_duplicates() that sequentially processes the elements in the linked list, using the variable CURR that is initially equal to the root node. During each iteration, the variable SEEK iterates through the successors to the CURR node: whenever there is a match, the latter node is deleted simply by skipping over that node, as shown here:

```
SEEK.next = SEEK.next.next
```

The final portion of Listing 6.7 initializes the arrays items, items2, and items3 to test the code with three different arrays of strings. Now launch the code in Listing 6.7 from the command line and you will see the following output:

```
=> list items:
Node: Stan
Node: Steve
Node: Stan
Node: George
Node: Stan

=> searching for duplicates
=> Found duplicate node # 1 with value: Stan
=> Found duplicate node # 2 with value: Stan
=> list items:
Node: Stan
Node: Steve
Node: George

=> searching for duplicates
=> list items:
Node: Stan
Node: Steve
Node: George

=> searching for duplicates
=> list items:
Node: Stan
Node: Steve
Node: George

=> searching for duplicates
=> list items:
Node: Stan
Node: Steve
Node: George

=> searching for duplicates
=> list items:
Node: Stan
Node: Steve
Node: George

original:
['Stan' 'Steve' 'Stan' 'George' 'Stan']
unique:
```

```
=> list items:
Node: Stan
Node: Steve
Node: George
```

TASK: CONCATENATE TWO LISTS

Listing 6.8 displays the contents of the Python file append_sllists.py that illustrates how to concatenate two linked lists.

LISTING 6.8: append_sllists.py

```
import numpy as np

class SLNode:
  def __init__(self, data):
    self.data = data
    self.next = None

def append_node(ROOT, LAST, item):
  if(ROOT == None):
    ROOT = SLNode(item)
    #print("1Node:", ROOT.data)
  else:
    if(ROOT.next == None):
      NEWNODE = SLNode(item)
      LAST = NEWNODE
      ROOT.next = LAST
      #print("2Node:", NEWNODE.data)
    else:
      NEWNODE = SLNode(item)
      LAST.next = NEWNODE
      LAST = NEWNODE
      #print("3Node:", NEWNODE.data)

  return ROOT, LAST

def display_items(ROOT):
  print("=> list items:")
  CURR = ROOT
  while(CURR != None):
    print("Node:", CURR.data)
    CURR = CURR.next
  print()

# append items to list1:
ROOT1 = None
LAST1 = None
```

```
#items1 = np.array([300, 50, 30, 80, 100, 200])
items1 = np.array([300, 50, 30])

for item in items1:
  ROOT1, LAST1 = append_node(ROOT1, LAST1, item)
  if(count == index):
    node2 = LAST1
  count += 1

display_items(ROOT1)

# append items to list2:
ROOT2 = None
LAST2 = None
#items2 = np.array([300, 50, 30, 80, 100])
items2 = np.array([80, 100, 200])

for item in items2:
  ROOT2, LAST2 = append_node(ROOT2, LAST2, item)
  if(count == index):
    node2 = LAST2
  count += 1

display_items(ROOT2)
```

```
# concatenate the two lists:
LAST1.next = ROOT2
display_items(ROOT1)
```

Listing 6.8 contains a function to create a linked list and another function that displays the contents of the the linked list. The next portion of Listing 6.8 initializes the NumPy arrays items1 and items2 and creates two linked lists with the contents of these two arrays.

The final portion of Listing 6.8 concatenates the two linked lists by setting the successor node of LAST1 (which is the last node in the first linked list) equal to the root node of the second linked list. Launch the code in Listing 6.8 and you will see the following output for the list of even length:

```
=> List of items:
Node: 300
Node: 50
Node: 30

=> List of items:
Node: 80
Node: 100
Node: 200

Now concatenate the two lists
=> List of items:
Node: 300
Node: 50
```

```
Node: 30
Node: 80
Node: 100
Node: 200
```

TASK: MERGE TWO LISTS

Listing 6.9 displays the contents of the Python file merge_sllists.py that illustrates how to merge two linked lists.

LISTING 6.9: merge_sslists.py

```python
import numpy as np

class SLNode:
  def __init__(self, data):
    self.data = data
    self.next = None

def append_node(ROOT, LAST, item):
  if(ROOT == None):
    ROOT = SLNode(item)
    #print("1Node:", ROOT.data)
  else:
    if(ROOT.next == None):
      NEWNODE = SLNode(item)
      LAST = NEWNODE
      ROOT.next = LAST
      #print("2Node:", NEWNODE.data)
    else:
      NEWNODE = SLNode(item)
      LAST.next = NEWNODE
      LAST = NEWNODE
      #print("3Node:", NEWNODE.data)

  return ROOT, LAST

def display_items(ROOT):
  print("=> list items:")
  CURR = ROOT
  while(CURR != None):
    print("Node:", CURR.data)
    CURR = CURR.next
  print()

# append items to list1:
ROOT1 = None
LAST1 = None
items1 = np.array([20, 75, 100, 150, 300])
for item in items1:
```

```
    ROOT1, LAST1 = append_node(ROOT1, LAST1, item)
    if(count == index):
      node2 = LAST1
    count += 1

display_items(ROOT1)

# append items to list2:
ROOT2 = None
LAST2 = None
items2 = np.array([80, 100, 200])
for item in items2:
  ROOT2, LAST2 = append_node(ROOT2, LAST2, item)
  if(count == index):
    node2 = LAST2
  count += 1

display_items(ROOT2)

# merge the two lists:
print("Merging the two lists:")
CURR1 = ROOT1
LAST1 = ROOT1
CURR2 = ROOT2
LAST2 = ROOT2
ROOT3 = None
LAST3 = None

#items1 = np.array([20, 300, 50, 30])
#items2 = np.array([80, 100, 200])

while(CURR1 != None and CURR2 != None):
  print("curr1.data:",CURR1.data)
  print("curr2.data:",CURR2.data)

  if(CURR1.data < CURR2.data):
    ROOT3, LAST3 = append_node(ROOT3, LAST3, CURR1.data)
    print("adding curr1.data:",CURR1.data)
    CURR1 = CURR1.next
  else:
    ROOT3, LAST3 = append_node(ROOT3, LAST3, CURR2.data)
    print("adding curr2.data:",CURR2.data)
    CURR2 = CURR2.next

# append any remaining elements of items1:
if(CURR1 != None):
  while(CURR1 != None):
    print("MORE curr1.data:",CURR1.data)
    ROOT3, LAST3 = append_node(ROOT3, LAST3, CURR1.data)
    CURR1 = CURR1.next

# append any remaining elements of items2:
if(CURR2 != None):
```

```
    while(CURR2 != None):
      print("MORE curr2.data:",CURR2.data)
      ROOT3, LAST3 = append_node(ROOT3, LAST3, CURR2.data)
      CURR2 = CURR2.next

# display the merged list:
display_items(ROOT3)
```

Listing 6.9 creates two linked lists from the two NumPy arrays items1 and items2. The second half of Listing 6.9 contains a loop that iterates through the elements of these two linked lists. During each iteration, the smaller value of the two values CURR1.data and CURR2.data is appended to the linked list items3. When the loop finishes execution, then either items1 is empty or items2 is empty: therefore, only one of the two subsequent loops will append "left over" elements from one of these lists to items3 (and it's also possible that *both* of these list are empty). The logic for this code is the same as the corresponding code sample in Chapter 5. Launch the code in Listing 6.9 and you will see the following output for the list of even length:

```
=> List of items:
Node: 20
Node: 75
Node: 100
Node: 150
Node: 300

=> List of items:
Node: 80
Node: 100
Node: 200

Merging the two lists:
curr1.data: 20
curr2.data: 80
adding curr1.data: 20
curr1.data: 75
curr2.data: 80
adding curr1.data: 75
curr1.data: 100
curr2.data: 80
adding curr2.data: 80
curr1.data: 100
curr2.data: 100
adding curr2.data: 100
curr1.data: 100
curr2.data: 200
adding curr1.data: 100
curr1.data: 150
curr2.data: 200
adding curr1.data: 150
curr1.data: 300
curr2.data: 200
```

```
adding curr2.data: 200
MORE curr1.data: 300
=> List of items:
Node: 20
Node: 75
Node: 80
Node: 100
Node: 100
Node: 150
Node: 200
Node: 300
```

TASK: SPLIT A SINGLE LIST INTO TWO LISTS

There are several ways to perform this task. One approach is to iterate through a given list and dynamically create a list of smaller items as well as a list of larger items in the loop. However, the logic is more complex, and therefore more error prone.

A simpler approach involves appending the smaller items to a Python list and then appending the remaining items to a larger list, and then return the two lists. At this point you can invoke the append() function to create two linked lists.

Listing 6.10 displays the contents of the Python file split_sllists.py that illustrates how to split a linked list into two lists.

LISTING 6.10: split_sllists.py

```
import numpy as np

class SLNode:
  def __init__(self, data):
    self.data = data
    self.next = None


def append_node(ROOT, LAST, item):
  if(ROOT == None):
    ROOT = SLNode(item)
    #print("1Node:", ROOT.data)
  else:
    if(ROOT.next == None):
      NEWNODE = SLNode(item)
      LAST = NEWNODE
      ROOT.next = LAST
      #print("2Node:", NEWNODE.data)
    else:
      NEWNODE = SLNode(item)
      LAST.next = NEWNODE
      LAST = NEWNODE
      #print("3Node:", NEWNODE.data)
```

```
      return ROOT, LAST
  def delete_node(node):
    found = False

    if(node != None):
      if(node.next != None):
        found = True
        print("curr node:",node.data)
        node.data = node.next.data
        node.next = node.next.next
        print("new  node:",node.data)

    if(found == False):
      print("* Item",node.data,"not in list *")

  def split_list(ROOT, value):
    node = ROOT
    smaller = list()
    larger  = list()

    while(node != None):
      if(node.data < value):
        print("LESS curr node:",node.data)
        smaller.append(node.data)
      else:
        print("GREATER curr node:",node.data)
        larger.append(node.data)
      node = node.next

    return smaller, larger

  def display_items(ROOT):
    print("=> list items:")
    CURR = ROOT
    while(CURR != None):
      print("Node:", CURR.data)
      CURR = CURR.next
    print()

  ROOT  = None
  LAST  = None

  # append items to list:
  items = np.array([10, 50, 30, 80, 100])
  for item in items:
    ROOT, LAST = append_node(ROOT, LAST, item)
    if(count == index):
      node2 = LAST
    count += 1
```

```
display_items(ROOT)
value = node2.data
smaller, larger = split_list(ROOT, value)
print("smaller list:",smaller)
print("larger  list:",larger)
```

Listing 6.10 starts with the definition of the Python class SLNode for elements in a singly linked list, followed by the methods append_node() and delete_node() that you have already seen in previous examples in this chapter.

The next portion of Listing 6.10 defines the split_list() method that creates a new singly linked list of elements whose values are smaller than the parameter value. Note that the comparison starts from the root node of an already constructed singly linked list. If a node contains a value that is larger than the variable value, then the node is appended to another singly linked list called larger. The final code snippet in Listing 6.10 returns the singly linked lists smaller and larger.

The next portion of Listing 6.10 defines the display_items() method that you have also seen in previous examples, followed by a code block that constructs an initial singly linked list of numbers. The final code block invokes the split_list() method and then displays the two singly linked lists that are returned by this method. Launch the code in Listing 6.10 and you will see the following output for the list of even length:

```
=> Initial list of items:
Node: 10
Node: 50
Node: 30
Node: 80
Node: 100

LESS curr node: 10
GREATER curr node: 50
GREATER curr node: 30
GREATER curr node: 80
GREATER curr node: 100
smaller list: [10]
larger  list: [50, 30, 80, 100]
```

TASK: FIND THE MIDDLE ELEMENT IN A LIST

One solution involves counting the number of elements in the list and then finding the middle element. However, this task has the following constraints:

- Counting the number of elements is not allowed
- No additional data structure can be used
- No element in the list can be modified or marked
- Lists of even length can have two middle elements

This task belongs to a set of tasks that use the same technique: one variable iterates sequentially through a list and a second variable iterates twice as quickly through the same list.

Listing 6.11 displays the contents of the Python file `middle_slnode.py` that determines the middle node in a singly linked list.

LISTING 6.11: middle_slnode.py

```
import numpy as np

class SLNode:
  def __init__(self, data):
    self.data = data
    self.next = None

def append_node(ROOT, LAST, item):
  if(ROOT == None):
    ROOT = SLNode(item)
    #print("1Node:", ROOT.data)
  else:
    if(ROOT.next == None):
      NEWNODE = SLNode(item)
      LAST = NEWNODE
      ROOT.next = LAST
      #print("2Node:", NEWNODE.data)
    else:
      NEWNODE = SLNode(item)
      LAST.next = NEWNODE
      LAST = NEWNODE
      #print("3Node:", NEWNODE.data)

  return ROOT, LAST

def find_middle(node):
  if(node == None):
    return None, -1
  elif(node.next == None):
    return node, 1

  MIDP = None
  CURR = node
  SEEK = node

  even_odd = -1
  while (SEEK != None):
    print("SEEK:",SEEK.data)
    if(SEEK == None):
      print("1break: null node")
      even_odd = 0
      break
    elif(SEEK.next == None):
      print("2break: null node")
      even_odd = 1
```

```
    elif(SEEK.next.next == None):
      print("3break: null node")
      even_odd = 0
    else:
      SEEK = SEEK.next.next
      CURR = CURR.next

    if(even_odd >= 0):
      print("returning middle:",CURR.data)
      return CURR, even_odd

ROOT  = None
LAST  = None

# append items to list:
items = np.array(["Stan", "Steve", "Sally", "Alex",
"Dave"])
#items = np.array(["Stan", "Steve", "Sally", "Alex"])
for item in items:
  ROOT, LAST = append_node(ROOT, LAST, item)

print()
print("=>items:")
print(items)

CURR = ROOT
while(CURR != None):
  print("Node:", CURR.data)
  CURR = CURR.next

middle,even_odd = find_middle(ROOT)

if(even_odd == 1):
  print("list has an odd  number of items")
else:
  print("list has an even number of items")
```

Listing 6.11 starts with the definition of the Python class SLNode for elements in a singly linked list, followed by the methods append_node() and delete_node() that you have already seen in previous examples in this chapter.

The next portion of Listing 6.11 defines the find_middle() method that will locate the middle element of a singly linked list. After handling a corner case, this method initializes the variables MIDP, CURR, and SEEK with the values None, node, and node, respectively. Note that the variable node is a parameter whose value is actually the ROOT node of a list that is constructed later in this code sample. In addition, the variable SEEK is the variable that will traverse the nodes of the singly linked list.

The next code block is a while loop that iterates through the singly linked list, as long as the variable SEEK is not null. The while loop performs conditional logic to check for the following three scenarios:

1. SEEK is None
2. SEEK.next is None
3. SEEK.next.next is None

The three preceding cases initialize the value of even_odd to 0, 1, and 0, respectively, which indicates whether or not the list has even length or odd length. Notice that the final else statement is where the variable SEEK is advanced *twice*, whereas the variable CURR is advanced only *once*. Therefore, when SEEK reaches the end of the singly linked list, the variable CURR will be at the middle of the linked list.

The next portion of Listing 6.11 constructs a singly linked list, followed by an invocation of the method find_middle() with the newly constructed singly linked list. If the value of even_odd is 1, the linked list has odd length, and if the value of even_odd is 0 then the list has even length.

Launch the code in Listing 6.11 and you will see the following output for the list of even length:

```
=> List of items:
['Stan' 'Steve' 'Sally' 'Alex']
Node: Stan
Node: Steve
Node: Sally
Node: Alex
SEEK: Stan
SEEK: Sally
3break: null node
returning middle: Steve
list has an even number of items
```

Now switch to the list of odd length in Listing 6.12 and when you launch the code you will see the following output for the list of odd length:

```
=> List of items:
['Stan' 'Steve' 'Sally' 'Alex' 'Dave']
Node: Stan
Node: Steve
Node: Sally
Node: Alex
Node: Dave
SEEK: Stan
SEEK: Sally
SEEK: Dave
2break: null node
returning middle: Sally
list has an odd  number of items
```

TASK: REVERSING A LINKED LIST

Listing 6.12 displays the contents of the Python file reverse_sllist.py that illustrates how to reverse the elements in a linked list.

_LISTING 6.12: reverse_sllist.py_

```
import numpy as np

class SLNode:
  def __init__(self, data):
    self.data = data
    self.next = None

def append_node(ROOT, LAST, item):
  if(ROOT == None):
    ROOT = SLNode(item)
    ROOT.next = ROOT
    LAST = ROOT
    print("1ROOT:", ROOT.data)
  else:
    if(ROOT.next == ROOT):
      NEWNODE = SLNode(item)
      LAST = NEWNODE
      ROOT.next = LAST
      LAST.next = ROOT
      print("2ROOT:", ROOT.data)
      print("2LAST:", LAST.data)
    else:
      NEWNODE = SLNode(item)
      NEWNODE.next = LAST.next
      LAST.next = NEWNODE
      LAST = NEWNODE
      print("3Node:", NEWNODE.data)

  return ROOT, LAST

def reverse_list(node, rev_list):
  if(node == None):
    print("1rev_list:",rev_list)
    return rev_list

  if (node.next == None):
    #print("2rev_list:",rev_list)
    return [node.data] + rev_list

  if (node.next != None):
    #print("3finding next node...")
    rev_list = [node.data] + rev_list
    #print("3rev_list:",rev_list)
    return reverse_list(node.next, rev_list)

ROOT  = None
LAST  = None

# append items to list:
items = np.array(["Stan", "Steve", "Sally", "Alex"])
for item in items:
```

```
    ROOT, LAST = append_node(ROOT, LAST, item)

# display items in list:
print("=> Original list items:")
CURR = ROOT
while(CURR != None):
  print("Node:", CURR.data)
  CURR = CURR.next
print()

print("=> Reversed list of items:")
rev_list = []
reversed = reverse_list(ROOT, rev_list)
print(reversed)
```

Listing 6.12 defines a Node class as before, followed by the Python function append_node() that contains the logic for initializing a singly linked list and also for appending nodes to that list.

Now launch the code in Listing 6.12 from the command line and you will see the following output:

```
=> Original list items:
Node: Stan
Node: Steve
Node: Sally
Node: Alex

=> Reversed list of items:
['Alex', 'Sally', 'Steve', 'Stan']
```

TASK: CHECK FOR PALINDROMES IN A LINKED LIST

A palindrome is a string (either numeric, character, or combination) that is the same as its reversed string. Examples of palindromes include 121, 1234321, radar, rotor, and so forth.

Listing 6.13 displays the contents of the Python file palindrome.py that illustrates how to determine whether or not a list contains a palindrome.

LISTING 6.13: palindrome.py

```
import numpy as np

class SLNode:
  def __init__(self, data):
    self.data = data
    self.next = None

def append_node(ROOT, LAST, item):
  if(ROOT == None):
    ROOT = SLNode(item)
    ROOT.next = ROOT
    LAST = ROOT
```

```
      print("1ROOT:", ROOT.data)
    else:
      if(ROOT.next == ROOT):
        NEWNODE = SLNode(item)
        LAST = NEWNODE
        ROOT.next = LAST
        LAST.next = ROOT
        print("2ROOT:", ROOT.data)
        print("2LAST:", LAST.data)
      else:
        NEWNODE = SLNode(item)
        NEWNODE.next = LAST.next
        LAST.next = NEWNODE
        LAST = NEWNODE
        print("3Node:", NEWNODE.data)

  return ROOT, LAST

def reverse_sum(node,sum,pow,base):
  if(node == None):
    print("1reverse: empty node")
    return sum

  if (node.next == None):
    print("2no child node value:",node.data)
    term = node.data * (base**pow)
    print("returning sum:",sum+term)
    return sum + term

  if (node.next != None):
    term = node.data * (base**pow)
    print("data:",node.data,"3finding next node...")
    return reverse_sum(node.next, sum+term, pow+1, base)
    print("4node value:",node.data,"sum:",sum,"pow:",pow)

ROOT  = None
LAST  = None

# append items to list:
items = np.array(["a", "b", "c", "b", "a"])
#items = np.array(["a", "b", "c", "b", "c"])
#items = str(1234321)

for item in items:
  ROOT, LAST = append_node(ROOT, LAST, item)

# display items in list:
print("=> Original list items:")
CURR = ROOT
while(CURR != None):
  print("Node:", CURR.data)
  CURR = CURR.next
print()

print("=> Original list of items:")
print(items)
print()
```

```
print("=> Reversed list of items:")
rev_list = []
reversed = reverse_list(ROOT, rev_list)
print(reversed)

same = True
for ndx in range(0,len(items)):
  if(items[ndx] != reversed[ndx]):
    same = False
    break

if(same == True):
  print("found a palindrome")
else:
  print("not a palindrome")
```

Listing 6.13 defines a `Node` class as before, followed by the `Python` function `append_node()` that contains the logic for initializing a singly linked list and also for appending nodes to that list.

Now launch the code in Listing 6.13 from the command line and you will see the following output:

```
=> Original list items:
Node: a
Node: b
Node: c
Node: b
Node: a

=> Original list of items:
['a' 'b' 'c' 'b' 'a']

=> Reversed list of items:
['a', 'b', 'c', 'b', 'a']
found a palindrome
```

SUMMARY

This chapter started with code samples for displaying the first k nodes in a list as well as the last k nodes of a list. Then you learned how to display the contents of a list in reverse order and how to remove duplicates.

In addition, you saw how to concatenate and merge two linked lists, and how to split a single linked list. Then you learned how to remove the middle element in a list and how to determine whether or not a linked list contains a loop.

Finally, you learned how to calculate the sum of the elements in a singly linked list and how to check for palindromes in a singly linked list.

## CHAPTER 7

QUEUES AND STACKS

This chapter introduces you to queues and stacks that were briefly introduced in Chapter 4. The first part of this chapter explains the concept of a queue, along with `Python` code samples that show you how to perform various operations on a queue. Some of the code samples also contain built-in functions for queues, such as `isEmpty()`, `isFull()`, `push()`, and `dequeue()`.

The second part of this chapter explains the concept of a stack, along with `Python` code samples that show you how to perform various operations on a stack. In addition, you will see code samples for finding the largest and smallest elements in a stack and reversing the contents of a stack.

The final section contains three interesting tasks that illustrate the usefulness of a stack data structure. The first task determines whether or not a string consists of well-balanced round parentheses, square brackets, and curly braces. The second task parses an arithmetic expression that can perform addition, subtraction, multiplication, or division, as well as any combination of these four arithmetic operations. The third task converts infix notation to postfix notation.

WHAT IS A QUEUE?

A *queue* consists of a collection of objects that uses the FIFO (first-in-first-out) rule for inserting and removing items. By way of analogy, consider a toll booth: the first vehicle that arrives is the first vehicle to pay the necessary toll and also the first vehicle to exit the tool booth. As another analogy, consider customers standing in a line (which in fact is a queue) in a bank: the person at the front of the queue is the first person to approach an available teller. The"back" of the queue is the person at the end of the line (i.e., the last person).

A queue has a maximum size MAX and a minimum size of 0. In fact, we can define a queue in terms of the following methods:

- isEmpty() returns True if the queue is empty
- isFull() returns True if the queue is full
- queueSize() returns the number of elements in the stack
- add(item) adds an element to the back of the queue if the queue is not full
- dequeue() removes the front element of the queue if the queue is not empty

In order to ensure that there is no overflow (too big) or underflow (too small), we must always invoke isEmpty() before "popping" an item from the top of the front of a queue and always invoke isFull() before "pushing" (appending) an item as the last element of a queue.

Types of Queues

The following list various types of queues that can be created, most of which are extensions of a generic queue, followed by a brief description:

- queue
- circular queue
- dequeue
- priority queue

A *queue* is a linear list that supports deletion from one end and insertion at the other end. A queue is a FIFO (first-in-first-out), just like a line of people waiting to enter a movie theater or a restaurant: the first person in line enters first, followed by the second person in line, and so forth. The term enqueue refers to adding an element to a queue, whereas dequeue refers to removing an element from a queue.

A *circular queue* is a linear list with the following constraint: the last element in the queue "points" to the first element in the queue. A circular queue is also called a *ring buffer*. By way of analogy, a conga line is a queue: if the person at the front of the queue is "connected" to the last person in the conga line, that is called a *circular queue*.

A *Dequeue* is a linear list that is also a double ended queue which insertions and deletions can be performed at *both* ends of the queue. In addition, there are two types of DQueues:

- Input restricted means that insertions occur only at one end.
- Output restricted means that deletions occur only at one end.

A *priority queue* is a queue that allows for removing and inserting items in any position of the queue. For example, the scheduler of the operating system of your desktop and laptop uses a priority queue to schedule programs for execution. Consequently, a higher priority task is executed before a lower priority task.

Moreover, after a priority queue is created, it's possible for a higher priority task to arrive: in this scenario, that new and higher priority task is inserted into the appropriate location in the queue for task execution. In fact, Unix has the so-called `nice` command that you can launch from the command line in order to lower the execution priority of tasks. Perform an online search for more information regarding the queues discussed in this section.

Now let's turn our attention to creating a basic queue along with some simple enhancements, which is the topic of the next several sections.

CREATING A QUEUE USING A PYTHON LIST

Listing 7.1 displays the contents of the Python file `myqueue.py` that illustrates how to use a `Python List` class in order to define `Python` functions to perform various operations on a queue.

LISTING 7.1: myqueue.py

```
import numpy as np

MAX = 4 # 100
myqueue = list()

def isEmpty():
  return len(myqueue) == 0

def isFull():
  return len(myqueue) == MAX

def dequeue():
  if myqueue:
    front = myqueue.pop(0)
    print("returning front:",front)
    return front
  else:
    print("* myqueue is empty *")

def push(item):
  if isFull() == False:
    myqueue.append(item)
  else:
    print("* myqueue is full *")

print("pushing values onto myqueue:")
push(10)
print("myqueue:",myqueue)
push(20)
print("myqueue:",myqueue)
```

```
push(200)
print("myqueue:",myqueue)
push(50)
print("myqueue:",myqueue)
push(-123)
print("myqueue:",myqueue)
print("myqueue:",myqueue)
print()

print("dequeue values from myqueue:")
dequeue()
print("myqueue:",myqueue)
dequeue()
print("myqueue:",myqueue)
dequeue()
print("myqueue:",myqueue)
dequeue()
print("myqueue:",myqueue)
dequeue()
print("myqueue:",myqueue)
dequeue()
print("myqueue:",myqueue)
```

Listing 7.1 starts by initializing myqueue as an empty list and assigning the value 4 to the variable MAX, which is the maximum number of elements that the queue can contain. (Obviously you can change this value.)

The next portion of Listing 7.1 defines several functions: the isEmpty function that returns True if the length of myqueue is 0 (and false otherwise), followed by the function isFull() that returns True if the length of myqueue is MAX (and False otherwise).

The next portion of Listing 7.1 defines the function dequeue that invokes the pop() method in order to remove the front element of myqueue, provided that myqueue is not empty. Next, the function push() invokes the append() method in order to add a new element to the end of myqueue, provided that myqueue is not full.

The final portion of Listing 7.1 invokes the push() function to append various numbers to myqueue, followed by multiple invocations of the dequeue() method to remove elements from the front of the queue. Launch the code in Listing 7.1 and you will see the following output:

```
pushing values onto myqueue:
myqueue: [10]
myqueue: [10, 20]
myqueue: [10, 20, 200]
myqueue: [10, 20, 200, 50]
* myqueue is full *
myqueue: [10, 20, 200, 50]
myqueue: [10, 20, 200, 50]

dequeue values from myqueue:
returning front: 10
myqueue: [20, 200, 50]
returning front: 20
myqueue: [200, 50]
```

```
returning front: 200
myqueue: [50]
returning front: 50
myqueue: []
* myqueue is empty *
myqueue: []
```

Listing 7.2 displays the contents of the Python file myqueue2.py that illustrates how to define a queue and perform various operations on the queue.

LISTING 7.2: myqueue2.py

```python
import numpy as np

MAX = 4 # 100
myqueue = list()

def isEmpty():
  return len(myqueue) == 0

def isFull():
   return len(myqueue) == MAX

def dequeue():
   if myqueue:
     front = myqueue.pop(0)
     print("returning front:",front)
     return front
   else:
     print("* myqueue is empty *")

def push(item):
   if isFull() == False:
     myqueue.append(item)
   else:
     print("* myqueue is full *")

arr1 = np.array([10,20,200,50,-123])

print("pushing values onto myqueue:")
for num in range(0,len(arr1)):
  push(num)
  print("myqueue:",myqueue)

print("dequeue values from myqueue:")
while(len(myqueue) > 0):
  dequeue()
  print("myqueue:",myqueue)
```

Listing 7.2 starts by initialing myqueue as an empty list and assigning the value 4 to the variable MAX, which is the maximum number of elements that the queue can contain. (Obviously you can change this value.)

The next portion of Listing 7.2 defines several functions: the isEmpty function that returns True if the length of myqueue is 0 (and False otherwise), followed by the function isFull that returns True if the length of myqueue is MAX (and False otherwise).

The next portion of Listing 7.2 defines the function dequeue90 that invokes the pop() method in order to remove the front element of myqueue, provided that myqueue is not empty. Next, the function push() invokes the append() method in order to add a new element to the back of myqueue, provided that myqueue is not full.

The final portion of Listing 7.2 invokes the push() function to append various numbers to myqueue, followed by multiple invocations of the dequeue() method to remove elements from the front of the queue. Launch the code in Listing 7.2 and you will see the same output at Listing 7.1.

CREATING A ROLLING QUEUE

Listing 7.3 appends and deletes elements from a queue, but we can make the code even simpler by combining a push and delete operation in the same function. Listing 7.3 displays the contents of the Python file rolling_queue.py that illustrates how to make sure that it's always possible to insert an element as the first element in a queue.

LISTING 7.3: rollingqueue.py

```
import numpy as np

MAX = 5 # maximum queue size
myqueue = list()

def isEmpty():
 return len(myqueue) == 0

def isFull():
   return len(myqueue) == MAX

def dequeue():
   if myqueue:
     front = myqueue.pop(0)
     print("returning front:",front)
     return front
   else:
     print("* myqueue is empty *")

def push(item):
   if isFull() == True:
     # remove last item:
     last_item = myqueue.pop()
     print("removed last item: ",last_item)
```

```
  # add new front item:
  myqueue.insert(0,item)
  print("new first item: ",item," queue: ",myqueue)

max = 100 # the number of elements for the queue
arr1 = [i for i in range(0,max)]

print("pushing values onto myqueue:")
for num in range(0,len(arr1)):
  push(num)
  #print("myqueue:",myqueue)

print("dequeue values from myqueue:")
while(isEmpty() == False):
  dequeue()
  print("myqueue:",myqueue)
```

Listing 7.3 is similar with Listing 7.2, along with a simple modification: if the queue is full, the push() method removes the final element of the queue and then inserts an element as the new first element of the queue. If need be, you can compare the code shown in bold in Listing 7.3 with the corresponding code in Listing 7.2. Now launch the code in Listing 7.3 and you will see the following output:

```
=> pushing values onto myqueue:
new first item:  0   queue:   [0]
new first item:  1   queue:   [1, 0]
new first item:  2   queue:   [2, 1, 0]
new first item:  3   queue:   [3, 2, 1, 0]
new first item:  4   queue:   [4, 3, 2, 1, 0]
removed last item:  0
new first item:  5   queue:   [5, 4, 3, 2, 1]
removed last item:  1
new first item:  6   queue:   [6, 5, 4, 3, 2]
removed last item:  2
new first item:  7   queue:   [7, 6, 5, 4, 3]
removed last item:  3
new first item:  8   queue:   [8, 7, 6, 5, 4]
removed last item:  4
new first item:  9   queue:   [9, 8, 7, 6, 5]
removed last item:  5
new first item:  10  queue:   [10, 9, 8, 7, 6]

// details omitted for brevity

new first item:  99  queue:   [99, 98, 97, 96, 95]
=> dequeue values from myqueue:
returning front: 99
myqueue: [98, 97, 96, 95]
returning front: 98
myqueue: [97, 96, 95]
returning front: 97
```

```
myqueue: [96, 95]
returning front: 96
myqueue: [95]
returning front: 95
myqueue: []
```

CREATING A QUEUE USING AN ARRAY

Listing 7.4 displays the contents of the Python file queue_array.py that illustrates how to use a Python List class in order to define a queue using an array.

LISTING 7.4: *queue_array.py*

```python
import numpy as np

MAX = 6 # 100
myqueue = [None] * MAX
print("myqueue:",myqueue)
print()
lpos = 2
rpos = 4

myqueue= 222
myqueue= 333
print("manually inserted two values:")
print("myqueue:",myqueue)

def isEmpty():
 return lpos == rpos

def isFull():
   return rpos >= MAX

def dequeue():
   global lpos,rpos
   if (lpos < rpos):
     front = myqueue[lpos]
     print("dequeued value:",front)
     myqueue[lpos] = None
     lpos += 1
     return front
   else:
     print("* myqueue is empty *")

def shift_left(myqueue):
   global lpos, rpos

   for i in range(0,rpos-lpos):
     myqueue[i] = myqueue[lpos+i]

   # replace right-most element with None:
   for i in range(rpos-lpos,rpos):
     #print("updating pos:",i)
     myqueue[i] = None
```

```
    print("Completed myqueue shift:",myqueue)
    rpos -= lpos
    lpos = 0
    return myqueue

def push(myqueue, item):
  global lpos, rpos

  if isFull() == False:
    print("rpos=",rpos,"pushing item onto myqueue:",item)
    myqueue[rpos] = item
    rpos += 1
  else:
    if(lpos == 0):
      print("*** myqueue is full: cannot push item:",item)
      print()
    else:
      print()
      print("Call shift_left to shift myqueue")
      print("before shift:",myqueue)
      print("left shift count:",lpos)
      myqueue = shift_left(myqueue)
      print("rpos=",rpos,"pushing item:",item)

      # now push the current item:
      print("rpos=",rpos,"Second try: pushing item onto
myqueue:",item)
      myqueue[rpos] = item
      rpos += 1
  return myqueue

arr1 = np.array([1000,2000,8000,5000,-1000])

print("=> Ready to push the following values onto
myqueue:")
print(arr1)
print()

for i in range(0,len(arr1)):
  myqueue = push(myqueue,arr1[i])
  if isFull() == False:
    print("appended",arr1[i],"to myqueue:",myqueue)

print("=> Ready to dequeue values from myqueue:")
while(lpos < rpos):
  dequeue()
  print("lpos:",lpos,"rpos:",rpos)
  print("popped myqueue:",myqueue)
```

Listing 7.4 starts by initializing the variables MAX (for the maximum size of the queue), myqueue (which is an array-based queue), along with the integers lpos and rpos that are the index positions of the first element and the last element, respectively, of the queue.

The next portion of Listing 7.4 defines the familiar functions isEmpty() and isFull() that you have seen in previous code samples. However, the dequeue() function has been modified to handle cases in which elements are popped from myqueue: each time this happens, the variable lpos is incremented by 1. Note that this code block is executed only when lpos is less than rpos: otherwise, the queue is empty.

The function shift_left is invoked when lpos is greater than 0 and rpos equals MAX: this scenario occurs when there are open "slots" at the front of the queue and the right-most element is occupied. This function shifts all the elements toward the front of the queue, thereby freeing up space so that more elements can be appended to myqueue. Keep in mind that every element in array is occupied when lpos equals 0 and rpos equals MAX, in which the only operation that we can perform is to remove an element from the front of the queue.

The final portion of Listing 7.4 initializes the NumPy array arr1 with a set of integers, followed by a loop that iterates through the elements of arr1 and invokes the push() function in order to append those elements to myqueue. When this loop finishes execution, another loop invokes the dequeue() function to remove elements from the front of the queue.

Change the value of MAX so that its value is less than, equal to, or greater than the number of elements in the array arr1. Doing so will exhibit different execution paths in the code. Note that in Listing 7.4, there are numerous print() statements that generate verbose output, thereby enabling you to see the sequence in which the code is executed (you can "comment out" those statements later). Now launch the code in Listing 7.4 and you will see the following output:

```
myqueue: [None, None, None, None, None, None]

manually inserted two values:
myqueue: [None, None, 222, 333, None, None]
=> Ready to push the following values onto myqueue:
[ 1000  2000  8000  5000 -1000]


rpos= 4 pushing item onto myqueue: 1000
appended 1000 to myqueue: [None, None, 222, 333, 1000, None]
rpos= 5 pushing item onto myqueue: 2000

Call shift_left to shift myqueue
before shift: [None, None, 222, 333, 1000, 2000]
left shift count: 2
Completed myqueue shift: [222, 333, 1000, 2000, None, None]
rpos= 4 pushing item: 8000
rpos= 4 Second try: pushing item onto myqueue: 8000
appended 8000 to myqueue: [222, 333, 1000, 2000, 8000, None]
rpos= 5 pushing item onto myqueue: 5000
*** myqueue is full: cannot push item: -1000

=> Ready to dequeue values from myqueue:
dequeued value: 222
```

```
lpos: 1 rpos: 6
popped myqueue: [None, 333, 1000, 2000, 8000, 5000]
dequeued value: 333
lpos: 2 rpos: 6
popped myqueue: [None, None, 1000, 2000, 8000, 5000]
dequeued value: 1000
lpos: 3 rpos: 6
popped myqueue: [None, None, None, 2000, 8000, 5000]
dequeued value: 2000
lpos: 4 rpos: 6
popped myqueue: [None, None, None, None, 8000, 5000]
dequeued value: 8000
lpos: 5 rpos: 6
popped myqueue: [None, None, None, None, None, 5000]
dequeued value: 5000
lpos: 6 rpos: 6
popped myqueue: [None, None, None, None, None, None]
```

This concludes the portion of the chapter pertaining to queues. The remainder of this chapter discusses the stack data structure, which is based on a LIFO structure instead of a FIFO structure of a queue.

WHAT IS A STACK?

In general terms, a stack consists of a collection of objects that follow the LIFO (last-in-first-out) principle. By contrast, a queue follows the FIFO (first-in-first-out) principle.

As a simple example, consider an elevator that has one entrance: the last person who enters the elevator is the first person who exits the elevator. Thus, the order in which people exit an elevator is the reverse of the order in which people enter an elevator.

Another analogy that might help you understand the concept of a stack is a stack of plates in a cafeteria:

- A plate can be added to the top of the stack if the stack is not full.
- A plate can be removed from the stack if the stack is not empty.

Based on the preceding observations, a stack has a maximum size MAX and a minimum size of 0.

Use Cases for Stacks

The following list contains use applications and use cases for stack-based data structures:

- recursion
- keeping track of function calls
- evaluation of expressions
- reversing characters
- servicing hardware interrupts
- solving combinatorial problems using backtracking

Operations With Stacks

Earlier in this chapter you saw `Python` functions to perform operations on queues; in an analogous fashion, we can define a stack in terms of the following methods:

- `isEmpty()` returns True if the stack is empty
- `isFull()` returns True if the stack is full
- `stackSize()` returns the number of elements in the stack
- `push(item)` adds an element to the "top" of the stack if the stack is not full
- `pop()` removes the top-most element of the stack if the stack is not empty

In order to ensure that there is no overflow (too big) or underflow (too small), we must always invoke `isEmpty()` before popping an item from the stack and always invoke `isFull()` before "pushing" an item onto the stack. The same methods (with different implementation details) are relevant when working with queues.

WORKING WITH STACKS

Listing 7.5 displays the contents of the `Python` file `mystack.py` that illustrates how to define a stack and perform various operations on the stack.

LISTING 7.5: mystack.py

```python
import numpy as np

MAX = 3 # 100
mystack = list()

def isEmpty():
 return len(mystack) == 0

def isFull():
   return len(mystack) == MAX

def pop():
   if len(mystack) > 0:
     top = mystack.pop()
     #print("returning top:",top)
     return top
   else:
     print("* mystack is empty *")

def push(item):
   if isFull() == False:
     mystack.append(item)
   else:
     print("* mystack is full *")
```

```
print("pushing values onto mystack:")
push(10)
print("mystack:",mystack)
push(20)
print("mystack:",mystack)
push(200)
print("mystack:",mystack)
push(-123)
print("mystack:",mystack)
push(50)
print("mystack:",mystack)
print()

print("popping values from mystack:")
pop()
print("mystack:",mystack)
pop()
print("mystack:",mystack)
pop()
print("mystack:",mystack)
pop()
print("mystack:",mystack)
pop()
print("mystack:",mystack)
```

Listing 7.5 is very similar to Listing 7.1, except that we are working with a stack instead of a queue. In particular, Listing 7.5 starts by initializing mystack as an empty list and assigning the value 3 to the variable MAX, which is the maximum number of elements that the stack can contain. (Obviously you can change this number.)

The next portion of Listing 7.5 defines several functions: the isEmpty function that returns True if the length of mystack is 0 (and False otherwise), followed by the function isFull that returns True if the length of mystack is MAX (and false otherwise).

The next portion of Listing 7.5 defines the function dequeue that invokes the pop() method in order to remove the front element of mystack, provided that mystack is not empty. Next, the function push() invokes the append() method in order to add a new element to the top of mystack, provided that myqueue is not full.

The final portion of Listing 7.5 invokes the push() function to append various numbers to mystack, followed by multiple invocations of the dequeue() method to remove elements from the top of mystack. Launch the code in Listing 7.5 and you will see the following output:

```
pushing values onto mystack:
mystack: [10]
mystack: [10, 20]
mystack: [10, 20, -123]
* mystack is full *
mystack: [10, 20, -123]
* mystack is full *
mystack: [10, 20, -123]
```

```
popping values from mystack:
mystack: [10, 20]
mystack: [10]
mystack: []
* mystack is empty *
mystack: []
* mystack is empty *
mystack: []
```

Listing 7.6 displays the contents of the Python file mystack2.py that illustrates how to define a stack and perform various operations on the stack.

LISTING 7.6: mystack2.py

```python
import numpy as np

MAX = 3 # 100
mystack = list()

def isEmpty():
 return len(mystack) == 0

def isFull():
  return len(mystack) == MAX

def pop():
  if len(mystack) > 0:
    top = mystack.pop()
    #print("returning top:",top)
    return top
  else:
    print("* mystack is empty *")

def push(item):
  if isFull() == False:
    mystack.append(item)
  else:
    print("* mystack is full *")

arr1 = np.array([10,20,-123,200,50])

print("pushing values onto mystack:")
for num in range(0,len(arr1)):
  push(num)
  print("mystack:",mystack)
print()

print("popping values from mystack:")
for num in range(0,len(arr1)):
  pop()
  print("mystack:",mystack)
print("mystack:",mystack)
```

Listing 7.6 is straightforward because it's a direct counterpart to Listing 7.2: the latter involves a queue whereas the former involves a stack. Launch the code in Listing 7.6 and you will see the following output:

```
pushing values onto mystack:
mystack: [0]
mystack: [0, 1]
mystack: [0, 1, 2]
* mystack is full *
mystack: [0, 1, 2]
* mystack is full *
mystack: [0, 1, 2]

popping values from mystack:
mystack: [0, 1]
mystack: [0]
mystack: []
* mystack is empty *
mystack: []
* mystack is empty *
mystack: []
```

TASK: REVERSE AND PRINT STACK VALUES

Listing 7.7 displays the contents of the `Python` file `reverse_stack.py` that illustrates how to define a stack and print its contents in reverse order. This code sample uses a `NumPy` array as a stack data structure.

LISTING 7.7: reverse_stack.py

```python
import numpy as np

MAX = 8 # 100
mystack = list()

def isEmpty():
 return len(mystack) == 0

def isFull():
  return len(mystack) == MAX

def pop():
  #print("len(mystack) =", len(mystack))
  if len(mystack) > 0:
    top = mystack.pop()
    #print("returning top:",top)
    return top
  else:
    print("* mystack is empty *")
    return None

def push(item):
  if isFull() == False:
```

```
    mystack.append(item)
  else:
    print("* mystack is full *")

arr1 = np.array([10,20,-123,200,50])

print("pushing values onto mystack:")
for i in range(0,len(arr1)):
  push(arr1[i])
  #print("mystack:",mystack)
print("mystack:",mystack)
print()

reversed = []
print("popping values from mystack:")
for num in range(0,len(arr1)):
  top = pop()
  reversed.append(top)
  #print("reversed:",reversed)
print("reversed:",reversed)
```

Listing 7.7 contains the code in Listing 7.6, along with a loop that invokes the push() function to insert the elements of the NumPy array arr1 (which contains integers) in the variable mystack.

After the preceding loop finishes execution, another loop iterates through the elements of mystack by invoking the pop() method, and in turn appends each element to the array reversed. As a result, the elements in the array reversed are the reverse order of the elements in mystack. Launch the code in Listing 7.7 and you will see the following output:

```
pushing values onto mystack:
mystack: [10, 20, -123, 200, 50]

popping values from mystack:
reversed: [50, 200, -123, 20, 10]
```

TASK: DISPLAY THE MIN AND MAX STACK VALUES (1)

Listing 7.8 displays the contents of the Python file stack_min_max.py that illustrates how to define a stack and perform various operations on the stack. This code sample uses a "regular" stack data structure.

LISTING 7.8: stack_min_max.py

```
import numpy as np

MAX = 6 # 100
mystack = [None] * MAX
lindex = 0
rindex = 0
min_val = np.Infinity
max_val = -np.Infinity
```

```python
def isEmpty():
 return lindex == rindex

def isFull():
  return rindex >= MAX

def pop():
  #print("len(mystack) =", len(mystack))
  if len(mystack) > 0:
    top = mystack.pop()
    #print("returning top:",top)
    return top
  else:
    print("* mystack is empty *")

def update_min_max_values(item):
    global min_val, max_val

    if(min_val > item):
       min_val = item

    if(max_val < item):
       max_val = item

def min():
  return min_val

def max():
  return max_val

def push(mystack, item):
  global rindex

  if isFull() == False:
    #print("1rindex=",rindex,"pushing item onto
mystack:",item)
    #mystack[rindex] = item
    mystack.append(item)
    rindex += 1
    #print("push 5rindex:",rindex)
    update_min_max_values(item)
  else:
    print("* mystack is full *")
    print("Cannot push item:",item)
    print("push 6rindex:",rindex)
  return mystack

arr1 = np.array([1000,2000,8000,5000,-1000])

print("pushing values onto mystack:")
for i in range(0,len(arr1)):
  mystack = push(mystack,arr1[i])
  print("mystack:",mystack)
print()

print("min value:",min_val)
print("max value:",max_val)
```

Listing 7.8 contains the familiar functions isEmpty(), isFull(), and pop() that have been discussed in previous code samples. Notice that the function pop() invokes the function update_min_max_values() each time that an element is removed from the stack. The latter method updates the variables min_val and max_val to keep track of the smallest and largest elements, respectively, in the stack. Launch the code in Listing 7.8 and you will see the following output:

```
=> Pushing list of values onto mystack:
[ 1000  2000  8000  5000 -1000]

min value: -1000
max value: 8000
```

CREATING TWO STACKS USING AN ARRAY

Listing 7.9 displays the contents of the Python file stack_array2.py that illustrates how to use an array in order to define two adjacent stacks and perform various operations on the stack.

LISTING 7.9: stack_array2.py

```python
import numpy as np

MAX = 6 # 100
mystack = [None] * MAX * 2
lindex1 = 0
rindex1 = 0
lindex2 = int(MAX/2)
rindex2 = int(MAX/2)

def isEmpty(num):
 if(num == 1):
   return lindex1 == rindex1
 else:
   return lindex2 == rindex2

def isFull(num):
  if(num == 1):
    return rindex1 >= int(MAX/2)
  else:
    return rindex2 >= int(MAX)

def pop(num):
  global lindex1,rindex1, lindex2,rindex2

  if(num == 1):
    if (lindex1 <= rindex1):
      print("pop position:",rindex1)
      rear = mystack[rindex1]
      print("popped value:",rear)
      mystack[rindex1] = None
```

```
          rindex1 -= 1
          return rear
      else:
        print("* mystack is empty *")
        return None
    else:
      if (lindex2 <= rindex2):
        print("pop position:",rindex2)
        rear = mystack[rindex2]
        print("popped value:",rear)
        mystack[rindex2] = None
        rindex2 -= 1
        return rear
      else:
        print("* mystack is empty *")
        return None

def push(mystack, num, item):
  global rindex1,rindex2

  if(num == 1):
    if isFull(num) == False:
      print("1rindex1=",rindex1,"pushing item onto
mystack:",item)
      mystack[rindex1] = item
      rindex1 += 1
      print("push 5rindex1:",rindex1)
    else:
      print("* mystack is full *")
      print("Cannot push item:",item)
      print("push 6rindex1:",rindex1)
    return mystack
  else:
    if isFull(num) == False:
      print("1rindex2=",rindex2,"pushing item onto
mystack:",item)
      mystack[rindex2] = item
      rindex2 += 1
      print("push 5rindex2:",rindex2)
    else:
      print("* mystack is full *")
      print("Cannot push item:",item)
      print("push 6rindex2:",rindex2)
    return mystack

print("=> Pushing list of values onto mystack:")
arr1 = np.array([1000,2000,8000,5000,-1000])
print(arr1)
print()

#print("A1index1:",lindex1,"rindex1:",rindex1)
#print("A1index2:",lindex2,"rindex2:",rindex2)

for i in range(0,len(arr1)):
  rand = np.random.rand()
  if(rand > 0.5):
```

```
      num = 1
    else:
      num = 2

  print("=> selected stack:",num)
  mystack = push(mystack,num, arr1[i])

#print("Blindex1:",lindex1,"rindex1:",rindex1)
#print("Blindex2:",lindex2,"rindex2:",rindex2)

print("------------------")
print("left stack:")
for idx in range(lindex1,rindex1):
  print(mystack[idx])
print()

print("right stack:")
for idx in range(lindex2,rindex2):
  print(mystack[idx])
print("------------------")
```

Listing 7.9 defines two stacks inside the stack `mystack` (which is an array): the left stack occupies the left half of `mystack` and the right stack occupies the right half of `mystack`. In addition, the variables `lindex1` and `rindex1` are the left-most and right-most index positions for the left stack, whereas the variables `lindex2` and `rindex2` are the left-most and right-most index positions for the right stack.

Notice that the usual functions `isEmpty()`, `isFull()`, and `push()` perform their respective operations based on the currently "active" stack, which is based on the value of the variable num: the values 1 and 2 correspond to the left stack and right stack, respectively.

One more difference is the loop that generates random numbers and then populates the two stacks based on whether or not each generated random number is greater than 0.5 (or not).

Launch the code in Listing 7.9 and you will see the following output:.

```
=> Pushing list of values onto mystack:
[ 1000   2000   8000   5000 -1000]

=> selected stack: 2
1rindex2= 3 pushing item onto mystack: 1000
push 5rindex2: 4
=> selected stack: 1
1rindex1= 0 pushing item onto mystack: 2000
push 5rindex1: 1
=> selected stack: 1
1rindex1= 1 pushing item onto mystack: 8000
push 5rindex1: 2
=> selected stack: 2
1rindex2= 4 pushing item onto mystack: 5000
push 5rindex2: 5
=> selected stack: 1
```

```
1rindex1= 2 pushing item onto mystack: -1000
push 5rindex1: 3
------------------
left stack:
2000
8000
-1000

right stack:
1000
5000
------------------
```

TASK: REVERSE A STRING USING A STACK

Listing 7.10 displays the contents of the Python file reverse_string.py that illustrates how to use a stack in order to reverse a string.

LISTING 7.10: reverse_string.py

```python
import numpy as np

MAX = 20 # 100
mystack = list()

def isEmpty():
 return len(mystack) == 0

def isFull():
   return len(mystack) == MAX

def pop():
   #print("len(mystack) =", len(mystack))
   if len(mystack) > 0:
     top = mystack.pop()
     #print("returning top:",top)
     return top
   else:
     print("* mystack is empty *")
     return None

def push(item):
   if isFull() == False:
     mystack.append(item)
   else:
     print("* mystack is full *")

my_str = "abcdxyz"

#print("pushing values onto mystack:")
for i in range(0,len(my_str)):
  push(my_str[I])
```

```
#print("mystack:",mystack)
#print()

reversed = ""
#print("popping values from mystack:")
for num in range(0,len(my_str)):
  top = pop()
  reversed += top
  #print("reversed:",reversed)

print("string:   ",my_str)
print("reversed:",reversed)
```

Listing 7.10 starts by initializing mystack as an empty list, followed by the usual functions isEmpty(), isFull(), and pop() that perform their respective operations. The next portion of Listing 7.10 initializes the variable my_str as a string of characters, and then pushes each character onto mystack. Next, a for loop removes each element from mystack and then appends each element to the string reversed (which initialized as an empty string). Launch the code in Listing 7.10 and you will see the following output:.

```
string:   abcdxyz
reversed: zyxdcba
```

TASK: BALANCED PARENTHESES

Listing 7.11 displays the contents of the file array_balanced_parens. py that illustrates how to use a NumPy array in order to determine whether or not a string contains balanced parentheses.

LISTING 7.11: array_balanced_parens.py

```
import numpy as np

def check_balanced(my_expr):
  left_chars = "([{"
  right_chars = ")]}"
  balan_pairs = np.array(["()","[]", "{}"])
  my_stack = np.array([])

  for idx in range(0,len(my_expr)):
    char = my_expr[idx]
    #print("char:",char)
    if char in left_chars:
      my_stack = np.append(my_stack,char)
      #print("appended to my_stack:",my_stack)
    elif char in right_chars:
      if(my_stack.size > 0):
        top = my_stack[len(my_stack)-1]
        two_chars = top + char
        #print("=> two_chars:",two_chars)
        if(two_chars in balan_pairs):
```

```
            #print("old stack:",my_stack)
            #remove right-most element:
            my_stack = my_stack[:-1]
            if(my_stack.size == 0):
              my_stack = np.array([])
            #print("new stack:",my_stack)
            continue
          else:
            #print("non-match:",char)
            break
        else:
          print("empty stack: invalid string",my_expr)
          return False
      else:
        print("invalid character:",char)
        return False

  return (my_stack.size == 0)

# main code starts here:
expr_list = np.array(["[()]{}{[()()]()}",
                      "[()]{}{[()()]()]",
                      "(()]{}{[()()]()]",
                      "(())(()()()()()]"])

for expr in expr_list:
  if( check_balanced(expr) == True):
    print("balanced string:",expr)
  else:
    print("unbalanced string:",expr)
```

Listing 7.11 is the longest code sample in this chapter that also reveals the usefulness of combining a stack with recursion in order to solve the task at hand, which is to determine which strings comprise balanced parentheses.

Listing 7.11 starts with the function check_balanced that takes a string called my_expr as its lone parameter. Notice the way that the following variables are initialized:

```
left_chars = "([{"
right_chars = ")]}"
balan_pairs = np.array(["()","[]", "{}"])
```

The variable left_chars and right_chars contain the left-side parentheses and right-side parentheses, respectively, that are permissible in a well-balanced string. Next, the variable balan_pairs is an array of three strings that represent a balanced pair of round parentheses, square parentheses, and curly parentheses, respectively.

The key idea for this code sample involves two actions and a logical comparison, as follows:

1. Whenever a *left* parenthesis is encountered in the current string, this parentheses is pushed onto a stack.

2. Whenever a *right* parenthesis is encountered in the current string, we check the top of the stack to see if it equals the corresponding left parenthesis.

3. If the comparison in #2 is true, we pop the top element of the stack.

4. If the comparison in #2 is false, the string is unbalanced.

Repeat the preceding sequence of steps until we reach the end of the string: if the stack is empty, the expression is balanced, otherwise the expression is unbalanced.

For example, suppose we the string my_expr is initialized as "()". The first character is "(", which is a left parenthesis: step #1 above tells us to push "(" onto our (initially empty) stack called mystack. The next character in my_expr is ")", which is a right parenthesis: step #2 above tells us to compare ")" with the element in the top of the stack, which is "(" . Since "(" and ")" constitute a balanced pair of parentheses, we pop the element "(" from the stack. We have also reached the end of my_expr, and since the stack is empty, we conclude that "()" is a balanced expression (which we knew already).

Now suppose that we the string my_expr is initialized as "((". The first character is "(" , which is a left parentheses, and step #1 above tells us to push "(" onto our (initially empty) stack called mystack. The next character in my_expr is "(", which is a left parenthesis: step #1 above tells us to push "(" onto the stack. We have reached the end of my_expr and since the stack is nonempty, we have determined that my_expr is unbalanced.

As a third example, suppose that we the string my_expr is initialized as "(()". The first character is "(", which is a left parentheses: step #1 above tells us to push "(" onto our (initially empty) stack called mystack. The next character in my_expr is "(", which is a left parenthesis: step #1 above tells us push another "(" onto the stack. The next character in my_expr is ")", and step #2 tells us to compare ")" with the element in the top of the stack, which is "(". Since "(" and ")" constitute a balanced pair of parentheses, we pop the topmost element of the stack. At this point we have reached the end of my_expr, and since the stack is nonempty, we know that my_expr is unbalanced.

Try tracing through the code with additional strings consisting of a sequence of parentheses. After doing so, the code details in Listing 7.11 will become much simpler to understand.

The next code block in Listing 7.11 initializes the variable my_expr as an array of strings, each of which consists of various parentheses (round, square, and curly). The next portion is a loop that iterates through the elements of my_expr and in turn invokes the function check_balanced to determine which ones (if any) comprise balanced parentheses. Launch the code in Listing 7.11 and you will see the following output:

```
balanced string:   [()]{}{[()()]()}
unbalanced string: [()]{}{[()()]()]
unbalanced string: (()]{}{[()()]()]
unbalanced string: (())(()()()()]
```

Consider enhancing Listing 7.11 so that the invalid character is displayed for strings that consists of unbalanced parentheses.

TASK: TOKENIZE ARITHMETIC EXPRESSIONS

The code sample in this section is a prelude to the task in the next section that involves parsing arithmetic expressions: in fact, the code in Listing 7.12 is included in the code in Listing 7.11. The rationale for the inclusion of a separate code sample is to enable you to tokenize expressions that might not be arithmetic expressions.

Listing 7.12 displays the contents of tokenize_expr.py that illustrates how to tokenize an arithmetic expression and also remove white spaces and tab characters.

LISTING 7.12: tokenize_expr.py

```
import shlex

try:
   from StringIO import StringIO
except ImportError:
   from io import StringIO

expr_list = ["            2789    *   3+7-8-   9",
             "4 /2   +   1",
             "2   -   3   +   4      "]

for expr in expr_list:
   input = StringIO(expr)
   result = list(shlex.shlex(input))
   print("string:",expr)
   print("result:",result)
```

Listing 7.12 starts with a try/except block that imports StringIO from the appropriate location, followed by the variable expr_list that contains a list of arithmetic expressions that include extra white spaces (including nonvisible tab characters).

The next portion of Listing 7.12 contains a loop that iterates through each element of expr_list and then invokes the shlex.shlex function that tokenizes each element. Launch the code in Listing 7.12 and you will see the following output:

```
string:        2789    *   3+7-8-   9
result: ['2789', '*', '3', '+', '7', '-', '8', '-', '9']
string: 4 /2   +   1
result: ['4', '/', '2', '+', '1']
string: 2   -   3   +   4
result: ['2', '-', '3', '+', '4']
```

TASK: EVALUATE ARITHMETIC EXPRESSIONS

Evaluating arithmetic expressions is an interesting task because there are various ways to approach this problem. First, some people would argue that the "real" way to solve this task involves lexers and parsers. However, the purpose of this task is to familiarize you with parsing strings, after which you will be better equipped to do so with nonarithmetic expressions.

Second, perhaps the simplest way involves the eval() function, which is a one-line solution and therefore the simplest solution. However, this solution does not familiarize you with parsing expressions.

After you have finished reading the code sample, you can enhance the code in several ways. For example, the operators "*" and "/" have equal priority, both of which have higher priority than "+" and "-" (and the latter pair have equal priority). The current code sample dos not take into account this priority, which means that "2+3*4" is evaluated as 20 (which is incorrect) instead of 14 (which is the correct answer). So, one variation involves adding the priority constraint for arithmetic operators.

The current code sample does not support round parentheses, square brackets, curly braces, or exponentiation (try adding these additional features after you finish reading this section).

Listing 7.13 displays the contents of the file parse_expr.py that illustrates how to parse and evaluate an arithmetic expression using a stack.

LISTING 7.13: parse_expr.py

```python
import numpy as np
import re

try:
  from StringIO import StringIO
except ImportError:
  from io import StringIO
import shlex

# performs "num1 oper num2"
# and returns the result
def reduction(num1,num2,oper):
  num1 = float(num1)
  num2 = float(num2)
  reduced = 0.0

  #print("RED string:",num1,oper,num2)
  if( oper == "*"):
    reduced = num1*num2
  elif( oper == "/"):
    reduced = num1/num2
  elif( oper == "+"):
    reduced = num1+num2
  elif( oper == "-"):
```

```
    reduced = num1-num2
  else:
    print("Binvalid operator:",oper)
  #print("returning reduced value:",reduced)
  return reduced

# a function that finds num1 and oper
# and reduces "num1 oper num2" and
# then puts this result on the stack
def reduce_stack(my_stack):
  num2 = my_stack[len(my_stack)-1]
  oper = my_stack[len(my_stack)-2]
  num1 = my_stack[len(my_stack)-3]
  #print("Anum1:",num1,"oper:",oper,"num2:",num2)

  # remove the right-most three elements:
  my_stack = my_stack[:-3]

  reduced = reduction(num1,num2,oper)
  my_stack = np.append(my_stack,reduced)
  #print("Creturning my_stack:",my_stack)
  return my_stack

# a function to place tokens on the stack
# and determine when to reduce the stack
def reduce_expr(my_expr):
  math_symbols = ["*","/","+","-"]
  #digits = [i for i in range(0,10)]
  digits = [str(i) for i in range(0,10)]
  my_stack = np.array([])
  oper = ""

  my_expr = strip_white_spaces(my_expr)
  for idx in range(0,len(my_expr)):
    token = my_expr[idx]
    if token in math_symbols:
      my_stack = np.append(my_stack,token)
      oper = token
    else:
      floatnum  = float(token)
      #print("found number in expr:",token)
      my_stack = np.append(my_stack,floatnum)
      if(oper != ""):
        my_stack = reduce_stack(my_stack)
      oper = ""

  return my_stack

# strip white spaces and tokenize symbols:
def strip_white_spaces(my_expr):
  expr2 = re.split(r'[ ]',my_expr)
  my_expr  = [token for token in expr2 if token != '']
  my_expr  = "".join(my_expr)
```

```
# tokenize string with symbols and no spaces (ex: '3+7-8-')
input = StringIO(my_expr)
new_expr = list(shlex.shlex(input))
#print("string:",my_expr)
#print("result:",new_expr)
return new_expr

expr_list = ["4 /2  +  1",
             "2  -  3  +  4  ",
             "        125  *  3+7-8-  9"]

for expr in expr_list:
  print("=> string:",expr)
  result = reduce_expr(expr)
  print("=> result:",result)
  print()
```

Listing 7.13 starts with a `try/except` block in order to import the Python `shlex` library, which will handle the "heavy lifting" in this code sample.

The next portion of Listing 7.13 contains the function `reduction`, that takes three parameters, where the first two are strings containing numbers and the third parameter is the arithmetic parameter to invoke on the first two parameters. After converting `num1` and `num2` to floating point numbers, an `if/elif` code block determine the value of `oper`, and then applies it to the other two parameters. For example, if `num1`, `num2`, and `oper` have the values 3, 4, and "*", the result is 3*4 = 12, which is returned to the calling function.

The next portion of Listing 7.13 contains the function `reduce_stack()` that takes a single parameter that is our current stack. This function pops the top three values from the stack and assigns them to `num2`, `oper`, and `num1`, respectively. Next, this function invokes the function `reduction()` to determine the result operating on the two numeric values, as shown here:

```
reduced = reduction(num1,num2,oper)
my_stack = np.append(my_stack,reduced)
```

As you can see, the purpose of this function is to perform a stack reduction operation. The next portion of Listing 7.13 contains the function `reduce_expr()` that starts by initializing the following variables:

- `math_symbols` consists of the four standard arithmetic operators
- `digits` is assigned the digits in the range of 0 to 9
- `my_stack` is initialized as an empty NumPy array
- `oper` is an empty string (and assigned something in math_symbols later)

The next portion of the function `reduce_expr()` initializes `my_expr` as the result of invoking the function `strip_white_spaces()`, which is where the "heavy lifting" is performed in this code sample.

The next section in `reduce_expr()` contains a loop that iterates through each character called `token` in the string `my_expr`, and performs the following logic:

- If `token` is a math symbol, append it to `my_stack` and set `oper` equal to token.
- Otherwise, append the floating point version of `token` to the stack, and if `oper` is not null, invoke the `reduce_stack()` function (described above).

When the preceding loop finishes execution, the function returns the updated contents of `my_stack` to the calling function.

The important function `strip_white_spaces()` removes redundant white spaces from `my_expr` and assigns the result to `expr2`, after which `expr2` is tokenized and then used to reinitialize the contents of `my_expr`. Then the `join()` operations concatenates all the elements of `my_expr`. At this point, we invoke `shlex()` that returns a perfectly parsed arithmetic expression.

The final portion of Listing 7.13 initializes the variable `expr_list` as an array of arithmetic expressions, followed by a loop that invokes the function `reduce_expr` with each element of `expr_list`, and then prints the evaluated expression. Now launch the code in Listing 7.13 and you will see the following output:

```
=> string: 4    /2 + 1
=> result: ['3.0']

=> string: 2 - 3 + 4
=> result: ['3.0']

=> string:          125 * 3+7-8-  9
=> result: ['365.0']
```

INFIX, PREFIX, AND POSTFIX NOTATIONS

There are three well-known and useful techniques for representing arithmetic expressions.

Infix notation involves specifying operators *between* their operands, which is the typical way that we write arithmetic expressions (example: 3+4*5).

Prefix notation (also called Polish notation) involves specifying operators *before* their operands, an example of which is here:

3+4*5 becomes + 3 * 4 5
3+4 becomes + 3 4

Postfix notation (also called Reverse Polish Notation) involves specifying operators *after* their operands, an example of which is here:

3+4*5 becomes 3 4 5 * +

The following table contains additional examples of expressions using infix, prefix, and postfix notation.

```
Infix          Prefix          Postfix
x+y            +xy             xy+
x-y            -xy             xy-
x/y            /xy             xy/
x*y            *xy             xy*
x^y            ^yx             yx^

(x+y)*z        *(x+y)z         (x+y)z*
(x+y)*z        *(+xy)z         (xy+)z*
```

Let's look at the following slightly more complex infix expression (note the "/" that is shown in bold):

```
[[x+(y/z)-d]^2]/(x+y)
```

We will perform an iterative sequence of steps to convert this infix expression to a prefix expression by applying the definition of infix notation to the top-level operator. In this example, the top-level operator is the "/" symbol that is shown in bold. We need to place this "/" symbol in the left-most position, as shown here (and notice the "^" symbol shown in bold):

```
/[[x+y/z-d]^2](x+y)
```

Now we need to place this "^" symbol immediately to the left of the second left square bracket, as shown here (and notice the "/" shown in bold):

```
/[^[x+(y/z)-d]2](+xy)
```

Now we need to place this "/" symbol immediately to the left of the variable y, as shown here (and notice the "+" shown in bold):

```
/[^[x+(/yz)-d]2](+xy)
```

Now we need to place this "+" symbol immediately to the left of the variable x, as shown here (and notice the "/" shown in bold):

```
/[^[+x(/yz)-d]2](+xy)
```

Now we need to place this "/" symbol immediately to the left of the variable x, as shown here, which is now an infix expression:

```
/[^[-(+(/yz))d]2](+xy)
```

The relative priority of arithmetic operators can be specified as follows:

```
precedence={'^':5,'*':4,'/':4,'+':3,'-':3,'(':2,')':1}
```

SUMMARY

This chapter started with an introduction to queues, along with real-world examples of queues. Next, you learned about several functions that are associated with a queue, such as `isEmpty()`, `isFull()`, `push()`, and `dequeue()`.

Next you learned about stacks, which are LIFO data structures, along with some `Python` code samples that showed you how to perform various operations on stacks. Some examples included reversing the contents of a stack and also included determining whether or not the contents of a stack formed a palindrome.

In the final portion of this chapter, you learned how to determine whether or not a string consists of well-balanced round parentheses, square brackets, and curly braces; how to parse an arithmetic expression; and also how to convert infix notation to postfix notation.

*I*NDEX